DIRT, SWEAT, AND DIESEL

DIRT, SWEAT, AND DIESEL

A Family Farm in the Twenty-first Century

STEVEN L. HILTY

UNIVERSITY OF MISSOURI PRESS
Columbia

Copyright © 2016 by
Steven L. Hilty
University of Missouri Press, Columbia, Missouri 65211
Printed and bound in the United States of America
All rights reserved. First printing, 2016.

ISBN: 978-0-8262-2079-0
Library of Congress Control Number: 2015960092

∞™ This paper meets the requirements of the
American National Standard for Permanence of Paper
for Printed Library Materials, Z39.48, 1984.

Typeface: Jenson

CONTENTS

Acknowledgments / vii
Introduction / 1

1 1971. The Beginning / 5

2 *December 5.* Snowstorm / 11

3 *December 12.* Seed Supper / 17

4 *December 16.* Maximilian / 29

5 *December 20.* Fences / 35

6 *January 9.* Help or Handout? / 47

7 *January 31.* Winter Work / 57

8 *February 16.* Mice and Mud / 67

9 *February 26.* Nursery / 77

10 *March 6.* Orphans / 87

11 *March 10.* Flood / 91

12 *April 1.* Spring / 101

13 *April 4.* Winter Tractor Pull / 113

14 *April 12.* Coming of Age / 121

15 *May 8.* Storm / 127

16 *May 18.* Thistles and Thieves / 139

17 *May 22.* Spring Planting / 149

18 *May 28.* Miracle Seeds? / 161

19 *June 11.* Fields of Fescue / 173

20 *June 15.* Community / 183

21 *June 25.* Breakdown / 189

22 *June 30.* Summer Harvest / 199

23 *July 4.* County Fair / 211

24 *July 24.* Prairie Tradition / 217

25 *August 15 & 18.* State Fair / 231

26 *August 25.* Summer Haze / 237

27 *September 12.* Waiting to Harvest / 249

28 *September 18.* Picking Corn / 255

29 *September 24.* Last Tractor Pull / 265

30 *September 27.* Milo / 275

31 *October 4.* Office with a View / 285

32 *October 8.* Farmfest / 297

33 *October 10.* Machinery Auction / 305

34 *October 14.* Video Auction / 315

35 *October 24.* Last Harvest: Soybeans / 319

36 *November 3.* Cattle Auction / 323

37 *November 10.* End of Innocence / 333

38 *November 17.* Little Bit Pregnant / 343

39 *November 30.* Cow on Fire / 353

Epilogue / 359

ACKNOWLEDGMENTS

All of the conversations and events in this book, and the descriptions of the land, people, farm life, and wildlife are real. They were obtained while working and visiting with the Montgomery family (not their real name) on their farm in western Missouri. Activities and conversations were recorded in notebooks on the spot or as soon thereafter as possible. In a few cases notes were dictated onto a minicassette recorder. Many events were photographed to preserve details for later descriptions. To protect privacy, the real names of people, including, neighbors, business associates and friends, however brief their appearance, have been changed. Most geographical and businesses names have been changed although real names have been retained for larger cities. I am deeply indebted to the families portrayed here for their hospitality and generosity, and for opening their homes and lives to me. They were unfailingly gracious, often going out of their way to teach me what I needed to know, to show me aspects of farming and community affairs that they felt were important and, through it all, went about their daily work unmindful that I was looking over their shoulders. Without their cooperation this book would not have been possible.

I also thank Harlan (not his real name), the family's full-time farm employee, for sharing many personal aspects of his life with humor and candor, as well as the many anonymous persons who appear in this book. They are uncelebrated heroes too, real people who live in the community and whose lives and work and personalities enrich us all.

This manuscript benefited greatly from the comments of two anonymous readers. Thanks also are extended to Joanne Asala, who edited the manuscript, and to the staff of the University of Missouri Press, including especially Gary A. Kass, Deanna Davis, Stephanie Williams, and David

Rosenbaum, as well as others at the press whose efforts, in various ways, helped to see this manuscript through to publication.

A Missouri Department of Conservation Natural Events Calendar was helpful in placing plant flowering and fruiting events and various wildlife activities in appropriate time sequences during the year.

Lastly, I thank my daughter Dru for encouraging this project after a college project piqued her interest in family farms, for reading and commenting on numerous chapters and for her insightful suggestions regarding a title. I also thank my daughter LaRae for accompanying me on several farm trips.

My greatest debt is to my wife Beverly, who read and commented on all of the chapters, some of them repeatedly, provided a stream of ideas and fresh perspectives, examined photos and illustrative material, and was always supportive.

I grew up on a family farm in Missouri but left to pursue another career. However, over the years, my family and I returned many times to this farm and continue to do so. This book is, in many ways, a product of my love of farming and farm life instilled in me by my parents.

INTRODUCTION

Once, more than three quarters of the American population counted themselves as rural, raising their families and earning their livelihoods on small family farms. They were part of a nation built on agriculture. Today, less than five percent of the population is rural, and less than one percent wrests a living from the land. In the last few decades, there has been renewed interest in rural life and, in some areas, rural populations have increased, but these people are mostly urban dwellers with urban jobs who have sought a less-stressful life in rural settings. In general, their livelihoods don't come entirely from the land, and it makes a difference. Those who depend on the land for their living, who roll their dice with weather and pestilence and uncertain commodity prices, see rural life differently than those who don't.

This is an account of the activities of a western Missouri farm family. It is a father and son partnership involving two households that live and work together. One wife works off the farm, the other worked off the farm part time for a few years. In both families incomes are derived almost entirely from the land they own and rent. Their farming business is varied and complex. The land they farm, about eighty-five percent of it rented, fed and clothed nine families a generation ago, and more than twice that many families two or three generations ago. Their farm operation continues to grow, as seems inevitable if family farms hope to survive in the increasingly tumultuous world of modern agribusiness. The men overcome a daunting and ever-changing array of technological hurdles in their daily business. They are not paid by the hour nor do they measure their financial successes and failures by counting the hours they work. Workdays are often long, but they accept the hours with stoicism—they wouldn't want it otherwise.

Both families are much more closely bound to modern technology than were earlier generations. Not surprisingly, they lack some of the independence that marked earlier rural generations when most meat, milk, vegetables, butter, and other foods were homegrown, and farm work was done with horses and simple equipment. They use computers in some aspects of their business but are not dependent upon them for day-to-day farm decisions. Rather, their success in farming stems more from a store of experience, an astute business acumen combined with frugality, and an intimate knowledge of cattle husbandry, farming basics, and of the mechanics of the machinery they use. They follow cattle, commodity, and farm machinery prices with an intensity that borders on obsession, attend several agriculture trade shows each year, and investigate anything that could potentially make their work easier or their business more profitable. No one requires their participation in these pursuits—it is what they do as successful farmers.

While rural and urban life are more similar today than in the past, some aspects of modern rural life still set it apart from that of most urban environments. Despite the vicarious violence that television and other forms of near-instant communication bring today, most urban dwellers are largely insulated from basic biological lifecycles, and from the connectedness to the land that farmers and ranchers feel. Urban dwellers see clean, well-organized fruit and vegetable displays and packaged meat on a supermarket shelf, milk in a carton, and butter in endless variety. Rural dwellers see the garden and orchard that battles weather and pests, the newborn calf from which the meat derives, and the cow that produces the milk or struggles to birth the calf. They see rats under grain bins, insects in hoppers of fresh-harvested grain, storms that wipe out a season's crop in minutes, newborn calves that freeze overnight or drown in floods, and sick cows that have to be put down. Cycles of life and death are part of the fabric of farm life, and it may seem that farmers are inured to them, but such cycles are taken in stride without moral judgment or reflection. Some readers may find this difficult to comprehend, and may find a few of the events described upsetting, but farmers generally don't entertain such judgments. Life continues, and those close to these cycles see themselves integrated within them rather than observers or evaluators apart from them.

Today, as our nation evolves away from its rural roots and away from the land that sustained and built this nation, the number of family farms is in decline. We may understand, in some general way, why this is happening, but that does not mean that it is desirable or that all family farms will disappear. The gradual loss of family farms and the inherent diversity that goes with them, however, leaves us as a nation increasingly vulnerable to forces beyond our control, and these are forces whose levers are increasingly operated by a small number of powerful agribusinesses whose interests may or may not align with those of consumers.

Most Americans today have likely never set foot on a modern farm, nor met a farmer, nor are likely to have much insight into daily life on a modern family farm. More often, what we learn about farm life is derived either from rather quaint accounts that conjure images of an era long past, or of accounts of futuristic, computerized techno-farms. Most family farms today are neither of these, instead lying somewhere between these extremes. What is true, however, is that in the past few generations, the craft of farming, and rural life generally, have changed dramatically.

These chapters provide a glimpse into the life and the mostly ordinary and unheralded work of a family that operates a mixed grain and cattle farm in western Missouri and, in a larger sense, into the community around it. The chapters are only snapshots—a scattering of days each month—but the events, the seasonal nature and variety of the work, family activities, joys, struggles, vagaries of weather, and eternal rhythms of land and seasons are real. So also are the seasonal changes in the land and the wildlife, which are seldom far from a farmer's eyes. Readers should note that while most material was gathered during 2005, a few activities were documented in 2006 and 2007 in an attempt to more fully complete an annual cycle of events in the lives of the men and women portrayed here. In the years since this material was gathered, their farming activities have not changed significantly, although their business continues to grow. The mix of crops grown each year varies somewhat in an attempt to take advantage of commodity price cycles, but the timelessness of seasonal cycles and the work itself remain unchanged.

1

1971
The Beginning

The land to the west of the Montgomery farmhouse is mostly flat, a panorama of fields and pastureland stitched together by tight, spare fences. From the homeplace there is an almost imperceptible downward slope for a mile or more until flat fields are finally interrupted by a single low hill that claims the sun's last rays. Del's grandfather had bought the farm after World War II, and for a time his parents had lived on the farm, the last of nearly a dozen farms his father had rented, or bought and sold. Eventually, Del's father moved to a nearby town to operate a grocery, but he continued to try to run the farm as well. For a while, Del's dad found a man to stay in the house on the farm and look after the cattle, but the tenant wasn't reliable. Then Del's grandfather moved back out to the farm, but by this time his health was failing and he wasn't able to do the hard physical labor. That was when Del stepped in to help. In his last year of high school, he was fresh from vocational agriculture classes that had honed his mechanical and welding skills and provided an introduction to agriculture basics. Del had an aptitude for mechanics that became evident even in high school when he tackled projects in his agriculture shop classes that were beyond the skills of his friends. But running a farm, even a small one, was a big responsibility for someone his age. He'd learned quite a bit about farming from helping his grandfather, and he convinced his dad that he could take over the farm. Flush with youthful optimism, he plunged into the work, planting, harvesting, and putting up hay. He bought a few cows. He also brushed aside the realities of a complicated and controversial war half a world away, even as military service remained obligatory. For two years after high school, Del lived his dream on that farm, and the experience he gained would prove invaluable.

Del had half expected the notice from the county draft board when it arrived. He had thought about trying to obtain an agricultural deferment. Maybe he could have gotten it, but in 1969 the military was bulking up in Vietnam. Beleaguered draft boards in rural counties with small populations were desperate for recruits. When the letter arrived, it was a jolt of reality. The lottery hadn't been established, so he reported for duty. Basic training was a nightmare, as it was for everybody, but the army, recognizing his mechanical aptitude, then trained him as a helicopter mechanic. Within weeks of his arrival in Vietnam, he was promoted to chief mechanic of a fleet of the army's Huey helicopters operating out of a remote base close to the North Vietnamese border. His year in Vietnam proved surprisingly uneventful. The base was rarely shelled. As he told a friend years later, "I had a clean, dry bed to sleep in every night, dry boots, and hot food. It was a lot better than the guys out in the jungle."

Looking back, he said it almost seemed like a normal job except for occasionally having to salute somebody. The South China Sea was a short distance away, and to relax he and his buddies made frequent excursions to the beach. But the site was isolated, and by the time his tour of duty was over he was ready to leave the war behind.

While Del was overseas, his grandfather died, and his father, who'd previously had a heart attack, struggled to manage the grocery and the farm. Del had been away for almost two years and he was anxious to get away from the war and helicopters. After his discharge, the first thing he did when he arrived home was drive out and look over the farm, walk the familiar fields of his youth, and start putting the chaos of war behind.

When he left the army, Del was offered several civilian opportunities in aircraft mechanics, but the farm beckoned. He felt the pull of open spaces, big skies, and his own land beneath his feet. He knew he would farm. It was destiny. He also knew he needed money. Even though he'd saved a little money in the army, it took a lot of capital to get started in farming, a lot more than a guy could save during a couple of years in the army, even with hazardous duty pay. The old equipment used by his dad and grandfather had to be replaced. He'd also met a girl that he wanted in his future. He was pretty sure she wanted to share that future, too, and he planned to talk to her about that right away.

One of the first problems Del faced was the old farmhouse. Built in 1886 from native lumber sawed locally, it was honeycombed with termite damage and the tin roof leaked. He immediately set to work tearing it down and replaced it with a doublewide trailer. It was his first major decision. He needed a place to live, and fast, because he had just talked to that special girl, and she had accepted his proposal.

Caitlin had grown up on a farm only ten miles west of the Montgomery homeplace, but she and Del had gone to different high schools and never met until later, partly by chance, when he was home on leave from the service. As a young girl on the farm, she had dreamed of living in a big city. It had seemed so exciting at the time, even though she and her little sister often played games where they pretended to be farmers or ranchers and rode stick horses. After high school, she started college and was halfway through when she met Del.

"At that point in my life, I never intended to live on a farm," she told friends. "I thought it was too much work. There were too many jobs I didn't like—plucking chickens and milking cows and cleaning manure out of barns. I thought the work was boring, and I wanted something different. That was why I started college. Then I fell in love with a man who wanted to farm, and ... when you marry a farmer, you marry his farm too."

Caitlin and Del got married a few months after he returned from Vietnam. Looking back, Caitlin said that she was surprised how quickly her attitude toward farming changed. She plunged into her new life with enthusiasm, helping whenever she could. Her farm background proved invaluable. Even jobs that she had once said she didn't like now took on new urgency. She drove tractors and trucks and raked hay on hot summer afternoons. She helped with winter chores and, a few years later, added taking care of their two small children to her growing list of activities. Raising children, especially when they were small, had been a bigger responsibility than she had imagined, and looking back, she wondered how she'd done it all. But even when the kids were little, she'd still managed to help with cattle roundups and vaccinations, and she made innumerable trips to sale barns to help buy and sell cattle, usually toting the kids along. During harvests, she trucked grain to elevators, her hair blowing in the wind of the open truck windows, her blue-gray eyes bright in the sun. She didn't like the big grain elevators because of the dust and noise, but she went anyway. It had to be done. Through it all she kept her slender, girlish figure and enjoyed dressing up on special occasions. It made her feel good after weeks of blue jeans and farm shirts, sunburned skin, and heavy farm boots.

When Del returned from the service, he and Caitlin both took jobs off the farm to help defray mounting farm expenses. Del also ramped up farm acreage by renting farmland. He bought a new tractor, an International Harvester, much larger than anything his grandfather had ever owned. It would be the only new tractor he would ever buy. A year later, he and Caitlin quit their day jobs. There was no turning back.

A small water-hauling business Del had started in the early years also helped. In the early 1970s, the county had no rural water system, and many farm families relied on cisterns for drinking water and for household use. Cisterns were typically hand-dug and notoriously unsanitary. They also were small and prone to going dry during droughts. Del and Caitlin's own marginal household water supply prompted him to begin hauling tanks of water from nearby Prairie Point. Then he hauled a few tanks of water as a favor for neighbors, but soon he found himself trucking thousand-gallon tanks of water to rural families all over the county. When he was busy, Caitlin drove the truck to town and filled the tank, and then Del made water deliveries in the evenings after farm work. He told Caitlin that some of the cisterns he saw were so dirty he wondered how people drank the water without getting sick. He'd seen spiders and frogs and whirligig beetles and all sorts of things in them, once even a little snake. He hauled water year-round. It was harder in the winter. Valves on the water tank froze, and he had to carry a blowtorch in the truck cab to thaw them. But the extra money had been nice. It paid for summer vacations, things the kids needed for school, even some farm expenses. Ten years after he began hauling water, a countywide rural water system was installed, but by then Del's farming operation had expanded.

Years later, remembering his water-hauling service, he told Caitlin, "It was a good thing the county got rural water. Otherwise I might still be driving that dang truck around and hauling water and ever'thing. We got enough to do without that."

Over time, Del and Caitlin acquired land of their own—the home forty acres, a forty across the blacktop road and a quarter mile to the east, an eighty four miles east of the homeplace, another eighty eleven miles west, and finally a hundred and sixty not far away. The forty across the road and the hundred and sixty he bought outright. The rest were family properties that he got by purchasing his brother's and sister's shares.

Del's grandfather had been a trader of horses and land. He also farmed, but neighbors said he traded horses so often that he never went to the field with the same team. Del's dad, also a trader and farmer, at one point acquired two houses in the town where he was a grocer, then traded one of the houses and some cash for an eighty-acre property. Del said his mother made a big fuss when she found out his dad had traded away one of the houses, and given three thousand dollars besides, for that eighty acres. She complained that the land was just rocks and timber. Told him he got skinned. At the time, Del remembered, it didn't seem like the best trade his dad could have made. Although only a mile east of town, the property was hilly and checkered with oak and hickory. A rocky stream that went dry in the summer crossed the property. It didn't compare with the flatter, richer prairie farmland west of town, but years later Del said the trade looked a lot better. With people from cities looking for wooded country property, it was worth as much as good farmland, maybe more, and the old house his dad had traded for it was now almost worthless.

There had been other land opportunities, but Del turned them down. He felt that farmland had gotten too expensive, especially in the 1980s. "Won't cash flow," he'd say. Other farmers, especially younger ones flush from a few years of good weather and crops, plunged during the 1970s and early 1980s. They purchased land at inflated prices, encouraged by bankers who should have known better. Many took on massive debt. Then a series of geopolitical decisions starting with Nixon's administration and continuing through the Carter and Reagan years went against American farmers. Commodity prices plummeted and stayed low. A lot of the plungers, along with their banks, failed. Newspapers were full of foreclosures. Del remained conservative, concentrating instead on renting farmland—a lot of it—and sharecropping rather than cash renting whenever possible. When he began farming, he took rental properties wherever he could find them. Some weren't nearby, but over time he let the distant ones go, and today most of the land he rents is within a few miles of the homeplace. Currently, nine landlords own the properties Del farms, about 2500 acres in addition to his own land. All nine properties had been single-family farms only a generation earlier. Today, few landowners live on these farms, and some rarely even visit the properties. It is a pattern repeated in many areas—a nationwide consolidation in agriculture and a flight to urban areas that began even before the

Great Depression. About 40% of the U.S. workforce was employed in agriculture in 1900. By the year 2000, it was only 1.9%. Paralleling this, farm size increased dramatically, as had labor efficiency from an estimated 27.5 acres/worker in 1980 to 740 acres/worker in 1990. By 2007 less than 1% of the U.S. population claimed farming as an occupation. In this regard the Montgomerys were an exception to this rural exodus. Starting from modest beginnings in the early 1970s and working steadily for four decades, they would today be regarded, by any measure, as successful farmers.

Most of the Montgomerys' farmland had originally been tallgrass prairie when settlers began moving into the area in the early 1800s. Biogeographers designate this west-central Missouri region as the Osage Plains. It is flat to gently rolling prairieland characterized by dark-brown, silty loam soils. Originally unglaciated, these prairies were mostly treeless and dominated by hardy prairie grasses. Eastward they give way to the Ozarks, a hilly region characterized by limestone outcrops and sturdy oak-hickory forests that cloak rugged valleys. The transition from prairie to Ozark forest is relatively abrupt, occurring in a curved band trending westward and then southwestward from near Columbia, Missouri, to the Oklahoma border.

The tallgrass prairies of North America once formed a belt of grassland bisecting the center of the continent and extending from central Canada to southern Texas. These prairies were the first extensive open areas encountered by settlers moving westward, and the rich soils beneath them would soon transform a young nation into the richest agricultural power on the planet. Today, tallgrass prairies, including those of Missouri, have mostly been ploughed and transformed into a rumpled patchwork of fields and farmsteads spilling westward across ever-flatter land. It is here, in west-central Missouri, along the prairies' eastern edge and not far from where the states of Missouri and Kansas meet, that the Montgomerys' ancestors settled more than a hundred and fifty years ago and put down farming roots.

Now Del had put down roots here too. He never wavered, never really doubted his ability to succeed, even in the face of a few money-losing years along the way. But farming was no guarantee of success. He often remarked that any farmer that has even survived the past few decades ought to consider himself a success. "Farming," he said, "is as much about surviving as making money."

2

December 5
Snowstorm

December. Winter cattle feeding begins; farm machinery repair occupies spare time in the farm shop; farmers purchase replacement cows and attend cattle sales; seed suppers held by seed companies; fence construction and repair possible until ground freezes; farmers sell stored grain if prices improve; families gather for holiday meals; year-end county and local taxes due; brush bulldozed as weather permits; barred owls call and court; Orion, Taurus, and other winter constellations prominent in evening sky.

It was still dark outside, but the snow reflected enough light for Del Montgomery to see the dim outlines of his farm buildings, the grain bins, even the corral fence. He stared out the back porch window. The half-buried root cellar behind the house was a snowy lump. Next to it he could see the silhouette of the old iron pitcher pump. One of the cats sleeping in a padded wicker chair on the porch stirred. Del pulled his phone from a pocket and called the owner of the cattle sale barn. It was early, but he'd be up. "How much snow you got up there?"

"They're saying almost a foot and a half," the voice at the other end answered. "Could be more. It's bad. We ain't even got down to the sale barn yet. Looks like we got to cancel the cattle sale tonight."

"Think you'll reschedule or just wait 'til next week?"

"Probably reschedule. A lot o' guys gonna be calling in the next hour or two. Buyers and sellers gotta know what's happening. Hell, we already got a few head of cattle brought in yesterday. Don't know what we're gonna do with them. If I was you, I'd tell them trucks you got coming to hold off."

"I'm about to do that. Trucks will just get stuck getting to the corrals. They'd be a mess to drag out of snowdrifts and ever'thing."

"We'll be giving you a call soon to straighten this deal out."

Del snapped his phone shut, a frown showing on his face. He turned and saw that Caitlin, his wife, had quietly joined him. Still in pajamas and robe, she was standing in the kitchen doorway.

"That takes care of the first crisis of the day," he said. "And it's not even first light yet."

Caitlin could see he was worried. "Come on inside. I'll get breakfast started. Then maybe things will look better," she said softly.

Del paused, then followed her into the kitchen. Caitlin had seen more than a few farm crises in her life. She was a good partner, steady, always supportive, even when things didn't go as they planned. She worried about things that affected the prosperity of the farm, too, but tried not to let it show.

The weather service had predicted the snowstorm. It wasn't a surprise, but it was one of the biggest early December storms Del could remember. Snow blanketed fescue fields and corn stubble where he was keeping his cattle at the moment. They'd be hungry. At this time of year, cows were already bred, most of them in their third trimester, eating for two. They looked nice—sleek and fat—but this was no time to have hungry cows walking around. For Del this snowstorm meant that winter chores had just begun.

The problem worrying Del right now was the cattle sale. One of his landlords was selling his herd. Del had arranged for two semis to load out the herd this morning, and he and his son, Zach, who farmed in partnership with him, had planned to haul the rest of the herd in their cattle trailers. Del guessed those cows weren't going anywhere for a few days. He pulled out his phone and called the trucking company.

Autumn harvests—corn, milo, and soybeans—were finished. Spring calves, most of them eight to nine months old and topping five hundred pounds, almost all had been sold by the end of November. Open cows, ones that hadn't bred, were already sold, too, and the money used to buy pregnant replacement cows. Del had purchased them at auctions. Fall was a busy time of the year in the cattle business—roundups, calf weanings, vaccinations, pregnancy tests, culling and selling old cows and nonbreeders, buying replacements, moving herds to fall pastures, and selling calves. Most of that was behind now. Usually there were a few weeks in late fall before the hard edge of winter arrived when cattle didn't require much attention. That ended last night.

Del liked cattle, said that they paid a lot of bills, even made a little money some years. They mostly took care of themselves during the summer, but winter was different. Winter cattle feeding usually began in December, after autumn grass, and after the cornstalks played out. Then the cattle had to be fed until spring grass. The February and March calving period was the busiest. When winter chores were light, Del and Zach and their hired man, Harlan, usually worked in the farm shop. Most shop work was routine maintenance of farm equipment. Sooner or later, everything came through the shop for servicing and repair, including teardowns, welding, cleaning, and even repainting. Many farm implements had a lot of moving parts, and if a part moved it eventually wore out. Del always said that some parts never seemed to do anything and still wore out. Shop work and looking after cattle mornings and evenings would comprise much of the winter routine.

By the time breakfast was finished, Del had his day planned, or hoped he did. Caitlin wondered out loud if she could get to her post office job with all the snow.

"Well, we're about to find out, hon," Del answered with a hint of resignation in his voice.

On the back porch he pulled on a pair of brown insulated coveralls, one of several that hung in a row on pegs along the west wall of the porch. Then he put on insulated rubber boots. At six feet and a hundred eighty pounds, Del was strong and fit, more energetic than his fifty-nine years would suggest. His hair was still thick and dark, his gray eyes steady. He had an easy smile, and when people met him it was always that smile and easygoing manner that people remembered. Years of sun had etched little crow's feet at the corners of his eyes. His face and muscled forearms were tanned most of the year, and his ubiquitous bill cap farmer haute couture. Seed companies, machinery dealers, feed stores, just about everybody in agribusiness handed out caps like calling cards. Del's cap-wearing habit, like that of most farmers, contributed to his ruddy face and onion-pale forehead, which was more visible on the infrequent occasions when he removed his cap. Most days Del wore old blue jeans, scuffed leather boots, and a cap. On cold days he added a hooded knit sweatshirt beneath insulated coveralls.

Zipping the legs of his coveralls, he straightened up, pushed against the porch door, which was partly blocked by snow, and stepped outside. The snow crunched and squeaked underfoot. Pebbly grains of sleet swirled in pirouettes in the gusty wind, peppering his boots and producing a swooshing sound like static when he walked.

Caitlin Montgomery was still in the kitchen when Del left. She heard the aluminum porch door slam and watched his dark form, bundled against the predawn cold, moving away across the driveway to the metal-covered barn where his feeding truck and a couple of tractors were stored. She smiled as she watched him, then pulled her long champagne-colored hair back and fastened it with a small elastic band. She and Del had been married for more than thirty-five years, and she couldn't remember them not getting up early. After breakfast he'd leave the house, often before first light, usually taking some food outside for the cats and the dog before starting farm work. Sometimes she went out with him, maybe to carry household trash to a burn barrel behind the farm shop, or to bottle-feed orphan calves during the winter calving season. For most of their married life, she'd been busy with farm work and raising their two children. She'd gone with Del and Zach to tractor pulls and farm shows and fairs, and driven trucks and tractors to help during busy times. When their daughter Wendy got interested in showing cattle, Caitlin helped her show at local and state fairs, even the prestigious American Royal in Kansas City. Now both children were grown and married, Zach farming with Del and Wendy working as a registered nurse.

Several years ago, Caitlin had gotten a job at the local post office. It was a part-time position, but it had changed her morning routine, and sometimes it almost seemed like full-time work. She delivered overnight and special-delivery packages and subbed for the letter carriers and the postmaster. She hadn't worked off the farm since her first year of marriage, wasn't used to wearing a uniform or working with computers or dealing with the record-keeping of the postal service, but she liked the varied hours and the excitement of the new job. There had been a big learning curve, and the hardest part, she said, had been learning everybody's job. Organizing the mail and driving a hundred-mile rural letter carrier route had taken time to learn.

This morning, Del was in a hurry. Keeping around three hundred and fifty cows healthy through the winter and through calving season was a big responsibility. He and Zach normally used their 4x4 pickup trucks to ferry bales of hay the size of Volkswagens to the herds. As he plodded across the driveway, leaving a trough in the snow, he had a feeling the trucks wouldn't be much use. He hadn't had time to put big mud and snow tires on the trucks yet. A tractor would be slower, but by using it he could be sure of getting out to his cattle.

Del was on his way to the first cowherd, one giant bale on the tractor's front loader, a second impaled on a spike behind, when he looked back toward the farmhouse. In the bluish half-light he saw a strew of exhaust vapor spiraling up behind Caitlin's truck as she was backing out of the semi-enclosed garage attached to the house. Partway out of the garage, the truck stopped. He watched the exhaust vapor pile up, a monstrous cloud. Then his phone rang.

"Yeah, hon, I'm on my way. I seen you're in trouble."

"I'll wait for you. Can't get through the snowdrift all piled up by the house. You have a tow chain?"

"Got two here in the tractor cab." Del left the two hay bales in the pasture for the first herd of cattle and then wheeled the big red tractor around toward the farmhouse. Now he was looking at his second crisis of the morning, and it wasn't even sunup yet. The county didn't have the money to plough all the roads, and graveled roads were out of the question. He figured that once he towed the truck out of the driveway, beyond the drifts, she would be fine. Del backed the tractor up behind the truck. On his knees and pushing away snow with one hand and arm, he reached underneath the truck to fasten a chain to the undercarriage.

"There's no tracks on the highway. Nobody's been out yet," Caitlin said.

"Not surprising, really," Del said, glancing toward the road. "Anybody with any sense would stay home this morning, unless they got to get to work."

"If you get me out to the blacktop, I'll be fine," Caitlin said. "I'll call if it looks bad. It's not that far."

"Town's probably got Main Street ploughed where the post office is."

"Wish me luck," Caitlin called back, then eased the truck forward.

Del watched her, figuring she'd be fine. On mornings like this he was reminded of why they drove four-wheel-drive trucks. When he and Caitlin were younger, they'd owned a couple of cars, then switched to trucks, which were more practical on the farm.

On more normal mornings, Caitlin might have fixed breakfast while Del relaxed or caught up on bills or listened to weather reports. But on days like this he couldn't relax. He had to know his cattle had food and water. He used to listen to market and farm commodity prices on the radio during breakfast, but most local stations didn't air live farm and commodity reports anymore. Now he usually just called Archer Daniels Midland's quote line in Memphis. For cattle prices he'd call a local auction barn. Phoning for quotes was quick, but it didn't provide the broad commodity picture that old radio farm reports did. He could get these prices on the Internet, but Del didn't like mouse clicking through pages of quotes. Besides, he'd often said that you could do other things while you listened to the radio, like eat breakfast. A computer, or even some little electronic tablet, just wasn't the same as a familiar voice quoting cattle futures.

Del figured there weren't enough farmers around anymore to justify radio stations airing long farm reports. He knew one station that still did, but it was too far away to pick up the signal most days. Caitlin didn't mind, however. She sometimes got tired of the long farm reports and would ask Del to change the station, but he never wanted to, afraid he'd miss something. She still teased him about his obsession with commodity prices, but he wasn't nostalgic for what once was. He said that when things changed you had to change too. Keeping up with commodity prices was essential, so he'd find other ways to stay up to date. To urban dwellers, rural life might seem rooted in the past, but farm life was changing just like everything else.

3

December 12
Seed Supper

A paved county road bordered the north side of the Montgomery homeplace. In the pasture west of the farm buildings, a small gulley originated near the highway, looped its way across the pasture, and continued southward beyond a tree-bordered fencerow. Most of the year the little gulley was dry.

On the opposite side of the homeplace there was another small pasture, and in the center of that pasture was an old hackberry tree that had escaped the blades of a mower years ago. In the summer, up to thirty black Brangus heifers, an Angus-Brahman breed hand-picked from a previous year's calf crop, usually grazed there. On hot summer days, these expectant heifers often cooled themselves in the shade of that gnarled old tree.

Across the gravel driveway from the house was a low, aluminum-covered barn with an open front. An extension on its north side contained a small pen for hand-reared baby calves. The barn also sheltered several trucks, an old Farmall M tractor, several restored antique and classic tractors, a portable feed grinder, assorted sacked mineral supplements and fertilizer, and extra rear tires for four large tractors. A white-painted cattle corral built from steel piping surrounded the north and west sides of the barn. The corral, a labyrinth of swinging gates and holding pens, had recently been partly covered with a metal roof to alleviate mud. At the front corner of the barn, the corral narrowed and led to a red metal squeeze chute where cattle were immobilized for vet work.

Behind the main house, there was a small white building with a gray roof. Originally built to smoke and cure meat butchered on the farm, it now served as storage. Next to it was a partly submerged root cellar. These were the only original structures remaining from homestead days. Beyond them lay Caitlin's flower and vegetable garden where dozens of perennials, herbs, old-fashioned hollyhocks, sunflowers, petunias, Spanish hats, and mints happily co-existed. Many spilled beyond their original boundary during the summer and some threatened to overtop the root cellar.

Behind the vegetable garden was the machine shop, a modern, metal-covered building trimmed in saddle brown. In many ways the shop was the work center of the farm. It brimmed with torches, welders, lathes, presses, compressors, drills, rows of tools and tools in rolling caddies, and an assortment of engines, engine parts, scrap steel, rows of shop bolts, nuts, washers, screws, belts, motor oil, transmission fluid, and grease buckets. A storage wing on the south side of the shop housed trucks and large machinery; a newer wing on the north side provided storage for more trucks and equipment. Beyond the shop there was a gleaming, reinforced steel Quonset building with a concrete floor, a building capable of holding twenty thousand bushels of grain. It was one of the newest buildings on the farm.

Across the driveway from the shop loomed eight matte-silver grain bins, each a giant, corrugated steel cylinder with a conical top like a rocket nosecone. Bulging full, they could hold some forty thousand bushels of grain. Electric grain dryers about the size of large household air conditioners protruded from the base of each bin. Del said that he couldn't control the weather but at least he could control storage and drying capacity, and that gave him control over prices for his harvested crops. Acquired over the years, they had been costly but had paid for themselves many times over. Most years the bins were filled before harvests ended, and sometimes emptied and refilled again. Beyond the steel grain bins a huge, sprawling, steel- and wood-framed machinery building housed his largest tractors and farm machinery.

In addition to their cattle, mostly Brangus and Black Angus, the Montgomerys maintained a growing inventory of machinery that included three tractor-trailer trucks and a fleet of trailers—gravity flow, gravel dump, flatbed, lowboy, two goose-neck flatbeds, two stock-trailers, a twin-axle pup trailer—and five pickup trucks, including three outfitted for handling large hay bales and one outfitted as a crop-spraying rig. They also owned three self-propelled combines. Their ever-shifting inventory of tractors, all of them International Harvester, included four large dual-wheeled tractors and several smaller tractors for light work, and more than a dozen others, including a competition pulling tractor and numerous restored antique and classic Farmall tractors, some dating to the 1930s. They also owned a bulldozer, two pull-type swathers, and an assortment of tandem disks, chisel ploughs, cultivators, sprayers, planters, mowers, brush-hogs, hay rakes, hay bailers, portable augers, a feed grinder, a tandem-axle truck with fertilizer-spreader bed and flotation tires, and a 1950s-era army "deuce and a half" truck. It was a massive equipment inventory and it would all be used during the course of a year's work.

This morning, clouds, like dark bruises, spread across a violet sky and outlined the old hackberry tree east of the house in stark relief. The snowstorm of a week ago was mostly gone now with only scattered patches of snow remaining in places shaded from the sun. Del loaded the skeleton of an old parted-out combine onto a flatbed trailer. Caitlin didn't like having junk equipment on the homeplace, so as soon as the engine had been removed and it had been stripped of everything else they thought might be useful—belts, bearings, chains, sieves, hydraulics—it would be consigned to a little corner of an eighty-acre plot they owned east of the town of Prairie Point. There, out of sight, it would join a metal scrap heap, stored until the price of iron rose and they could sell it. Until then it would shelter opportunistic oak and cedar and blackberry sprouts and an occasional cottontail rabbit or coyote wandering the property.

Everything Del and Zach salvaged from the old combine would fit at least one or two of their three newer combines. Del's strategy of using second-hand equipment and repairing it with low-cost used parts kept maintenance expenses minimal. With such extreme cost-efficiency, he was able to turn a profit most years, even when crops and weather conditions were marginal and other farmers were losing money.

In addition to cattle, their farming operation, and shop work, Zach also contributed to the farm partnership by taking on occasional bulldozing jobs for hire. Like many items in their inventory, their dozer was one Del had bought for scrap iron price. They'd repaired its leaking hydraulic cylinders and pumps, rebuilt the tracks and transmission, and then replaced the motor with a larger one that required adding eight inches of length to the engine compartment and to the dozer blade arms. It was a big job, but the retrofit worked perfectly, and the dozer had paid for itself many times over.

This morning Zach had a dozing job but he needed diesel fuel. Del usually didn't keep much diesel fuel around in the winter, so with the dozer on a lowboy trailer, he filled the dozer's tank at a highway truck stop in

Prairie Point while Zach fed cattle. It cost nearly four hundred dollars to fill the tank, and Del carefully folded the receipt in his wallet when he left the station. Farm and construction equipment, unlike vehicles on public roads, were eligible to burn a type of diesel fuel dyed red that was not taxed by the state, but it had to be delivered in bulk to farm tanks, and Del didn't want a huge tank of diesel delivered in the dead of winter when he had little need for it. Besides, it didn't store well when it was cold. So this morning he'd paid full price but kept the receipt. He could claim the tax credit later.

Farm gasoline was the opposite. Farmers paid tax on it, even when delivered in bulk, then had to request a tax rebate from the state so they could deduct state and federal portions of it on their income taxes. It was a bookkeeping nightmare. It also was tempting to use the red diesel in pickup trucks, something not permitted if the vehicle was used on highways, but the department of transportation wasn't asleep to this possibility. Occasionally they set up sting operations at sale barns and auctions and checked fuel tanks. Del had heard of a few guys that had got caught with red diesel in their pickup trucks, and he knew they'd written some pretty big checks for their indiscretions.

Harlan was Del's only employee. He'd worked as a hired hand for Del for over twelve years, and Del kept him busy even during the winter. Harlan had worked a lot of rough jobs in the past, mostly in the Midwest and south to the Texas oil patch. His often unshaven face was weathered and spare from working outside most of his life. He had nervous, dark-blue eyes and brownish, untamed hair that escaped in long curls from underneath his cap. He'd never held a job too long before moving, but after he started working for Del, he got married and settled down some. Farm work seemed to suit him, and under Del's guidance, he'd become a good mechanic. Some of the work was basic—changing oil and filters, replacing bearings, frayed belts, worn chains and sprockets, and pressure washing farm equipment—but he also handled engine and transmission teardowns and other complex tasks. Keeping Harlan around full time was expensive, but Del figured he could justify the cost. Del tried to keep all of his equipment in good working condition, but because he used older equipment it required more maintenance, and that was where Harlan came in. Preventing breakdowns and lost time during critical planting or harvesting seasons could mean the difference between a profit and a loss.

After his trip for diesel, Del drove back to the farm shop. Boy, the gray shop cat, was stretched out on top of a tall tractor tire, his lithe feline body and black barring conforming to the irregularity of the heavy lugs on the tire. His tail flicked in recognition as Del entered the shop. Boy had never really received a proper name like the other farm cats, and no one could recall when or why they started calling him Boy. It didn't seem to matter. He'd been there since he was a kitten and he'd been a permanent fixture in the farm shop, rarely leaving except on the coldest winter nights when he occasionally moved to the relative comfort of the farmhouse back porch.

"How's it look?" Del asked.

"Not so good. Feel that worn groove there?" Harlan said, rubbing a greasy finger along the inside of an exposed engine cylinder.

"Yeah, it's an oil burner."

"I reckoned it would be. What you want to do about it now that it's all tore down?"

"Go ahead and take this engine out and put the original motor back in, that is if you can find all the pieces."

"Isn't this the tractor you got from that machinery dealer?" Harlan asked.

"Yep. Engine was all torn down when we got it. It's still in those boxes there in the back of the shop. Dealer told me an old man brought the tractor in and wanted it overhauled. Came back with his wife a few days later. She argued with the dealer. Said the overhaul was too expensive. Dealer told her he already had several hundred dollars' worth of labor in the teardown and wanted to be paid for the labor. He offered to haul the tractor back to the farm but said he couldn't reassemble it, leastwise not all of it."

"Ain't that something?" Harlan said as he lit a cigarette. "Not very accommodating if you ask me."

"His wife got angry," Del continued. "She wouldn't pay, wouldn't let her husband pay neither. She told the dealer he could keep that old tractor. Said it wasn't worth the cost to fix it anyway, which was probably true considering what dealers charge for labor today. Truth is, she probably said a lot more than that to the dealer. I imagine she reamed him out good. Course, he didn't say that to me."

"Yeah, I bet she did. So what happened?"

"They just left it. Sat there for several months in the dealer's shop. He didn't want it neither. Wanted to get it out of the way. Then Zach and I heard about it. We offered him the cost of his labor for it just like it was. He was glad to get it off his hands. When we picked it up, the engine was all torn down, pistons and connecting rods and boxes of nuts and bolts all over the place."

"So how come it's been here so long?" Harlan asked. "I seen all those pieces of it sitting over there in boxes in the back of your shop for over a year now. Never knew where it came from."

"I think you were off in a fishing tournament or something when we got it. Anyway, it was summer, and we were busy and didn't have time to put the engine back together, but we thought we could use the tractor 'cause it had three sets of hydraulic pumps."

"Not many old tractors have any hydraulics, much less three sets."

"That's right. So we just put this old Farmall engine in it. It's the same as the one that's in pieces back there, but it's an oil burner, so we figured it was time to get rid of it and put the original engine back in this tractor."

"Have you looked at it?"

"We did, sort of. Couldn't see much wrong with it except the carburetor and fuel lines were all gummed up. We boiled them out not long ago, and I think they're okay. All you got to do is pull this old oil-burning motor out and put the original one back together and put it in this tractor."

"So, you're saying the original engine didn't need an overhaul?"

"Yeah. I think it's okay. Sad thing is, that old man probably lost his tractor for nothing. Makes you wonder about the mechanics working for that dealer. They ought to have figured that out right away."

"Hey! That Caitlin's truck I hear outside?" Harlan said, cocking his head.

"Sounds like she's home for dinner. On break from the post office."

Minutes later, Zach arrived in his old brown pickup, and the three of them crowded onto the back porch, pulling off boots and coveralls. Gathering around the circular kitchen table, each of them began setting out plates and food, moving chairs, and bumping into each other in the kitchen. Del put some ice cubes in glasses, then flooded one with too much tea, some of it splashing on the floor.

"Well," Caitlin teased, "you can fix on a motor or a transmission but you can't even pour a glass of tea without spilling it."

"Don't have no room to do anything in here," Del said a little defensively. "Besides, Harlan and Zach were in my way."

"I think you better stick to mechanics, Dad. Leave the kitchen to Mom," Zach said, laughing as he sat down beside Harlan.

After dinner, with Zach bulldozing and Harlan in the shop, Del towed a stock trailer to storage in a barn on a rental farm. He opened a swinging red metal gate, drove into a large corral, and angled the truck to the side to line up the trailer. Using the truck mirrors for guidance, he backed the long, double-axle trailer into the narrow slot, easing it into position on the first try. There was four inches of clearance on one side, barely six on the other.

After unhooking the trailer, he drove into a nearby pasture to get a headcount of the cows there. Within minutes, an audience of attentive cows gathered, rushing the truck like teenagers after a rock star, pushing against each other, bumping the truck, breath noisy and steamy, thick gray tongues licking wet noses in anticipation of a mineral supplement handout.

"Sorry, gals," Del grinned, "nothing today."

Del usually tried to get a headcount of his herds about once a week. Sometimes he did it on Sunday mornings if Caitlin went to church. That was when he'd take time to look over the herds and give them mineral cubes. Satisfied with his count, he checked two other herds in nearby pastures. On the way home, he passed several farmhouses but didn't see anyone outside. Roadside ditches were prickly with old, dry skeletons of giant ragweeds and sunflowers from the previous summer. Lines of brush in fencerows formed irregular borders to some of the fields, others were bare and stark. The land would look like this until spring.

Back home, Del checked on Harlan again. By now he had the little oil-burning engine unbolted from the tractor frame and a hoist attached to it.

"I'm ready to pull this motor, but I'll just leave it partly tore down until I find out if Zach wants to do something with it," Harlan said as he adjusted his cap with a grease-stained hand. "He may want to bore the cylinders or something and use it in Amber's pulling tractor."

"He might. Don't know what he has in mind, but Amber likes tractor pulling. You know, she'd never driven a tractor until she married him. Now she's in pulling contests."

"Seems surprising, her not being from a farm background and all."

"It does, but she even entered a pulling contest a couple weeks after Nathaniel was born."

"Did Zach show her what to do? You know, like driving and everything?"

"He did. He was surprised and at first he didn't think she was serious. Then we got an old Farmall H for her. Fixed it up a little. She said it suited her fine. She loves to compete against the men drivers. Occasionally, she even wins."

"We got sleeve inserts for this engine right here, but if Zach bores the block he'll have to get bigger ones," Harlan continued. "Anyway, I got the cherry picker already hooked up to here. Engine's ready to pull. Sleeves will stop the oil burning."

"Yeah, sort of like coronary surgery for old tractor motors, I guess," Del added as he turned his attention to his big red International 986 tractor sitting alongside the little Farmall. The torque converter, a shifting mechanism that allowed an operator to downshift a gear without clutching, wasn't working. Fixing it wasn't near as difficult as getting to it. He'd have to split the tractor in half, separating the front from the rear where the clutch, torque converter, and transmission were located. Del removed the batteries, battery holders, cab stairs, exhaust pipe, and engine cowling. Then, crawling under the tractor's cab, he removed the clutch and brake linkages, and that was just the beginning. He was sentimental about this tractor. He'd bought it in 1979, the only new tractor he'd ever owned. It had done thousands of hours of grueling fieldwork but was showing its age.

Del's cellphone rang. He dug through a couple of layers of jackets and coveralls to get to the phone. "Yep," he answered, continuing to loosen a nut beneath the tractor's cab.

"Dad, I finished dozing, cleaning out a pond on Larousse's place, but I got to clean up the dozer some before loading it. Got clay all over it. I'll be a little late picking you and Mom up for the seed supper."

"No problem. It's nearly six, and Caitlin just got off work at the post office, so we're not ready anyway. She said Christmas mail overwhelmed them. They couldn't finish on time."

Seed suppers were an annual affair in rural midwestern communities. Farmers swapped an evening listening to seed company promotions for a free meal and some socializing with neighbors. Seed companies mostly held suppers in late fall or early winter, after harvest, when farmers had a little free time. Del and Caitlin usually went to several. Everybody knew it was about promotion, but farmers had to buy their seed somewhere. If they could learn a little at the suppers—hear about the newest seed hybrids, what was working best in their area—so much the better.

Big multinational seed companies spent a lot on research to develop new lines of seeds, and they wanted the farmers' business. Farmers, for their part, spent heavily on seed purchases, so there was competition for their dollars when they went shopping. Gone were the days when a farmer simply held back a few sacks of seed from his current crop for next planting season. Seeds today were usually hybrids, and some of them wouldn't germinate even if they were held back, and others were genetically modified and covered by a barrage of patents and couldn't be saved back legally. A few guys had done it anyway, but Del had heard that when they got caught, they'd been sued by big seed companies like Monsanto, who aimed to protect their products. Roundup Ready soybeans were a prime example. Monsanto was the first company to develop Roundup Ready crops like soybeans, and you couldn't save back the seeds to plant the next year. Sometimes their tactics seemed awfully aggressive to Del, but what concerned him more were the high prices he had to pay for these kinds of seeds, and for the herbicides. He understood that everybody had to make a profit, but a lot of these new seeds were so expensive it was squeezing farmers. Del would spend sixty or seventy thousand dollars, maybe more, on corn and soybean seed in the coming year—part of that from the company hosting the supper tonight. He'd also buy seed directly from one or more other seed companies, and from the local farmers' co-op too. And he'd also have to buy milo seed, and in some years he also planted wheat. And every four or five years he usually had to replant some of the fescue grass. So, like most of the farmers that would be at the supper tonight, he was hardly a disinterested party just looking for a free meal.

When the Montgomerys arrived, several dozen farmers and their wives were already seated around a large U-shaped arrangement of banquet tables at the town meeting hall in Ashton. More couples arrived, eventually filling nearly every folding chair in the room. The men and women were in a good mood, and it was noisy. Most of the farmers knew each other. They joked and exchanged stories over coffee and tea, and found something to complain about or poke fun at.

It was two hours past sundown, but most of the men at the seed supper still wore caps, doffing them momentarily for the brief prayer that preceded the dinner. Some seemed older without their caps, their balding heads, thinning hair, and pale foreheads visible. Their caps, like the ones Del wore, were mobile advertisements for local machinery dealers, tools, fertilizers, seeds, and farm products the men used. Dress was casual. A few wore bib overalls, the rest jeans. Most men ate with their caps on. Their wives were considerably better dressed, but fashion wasn't uppermost in anyone's mind.

These men worked hard, took large financial risks with land and machinery and crops, and faced physical risks working around machinery and livestock. They took whatever the weather and markets and things beyond their control dished out with a certain amount of stoicism. But their conversations revealed that many of these issues were on their minds.

"I went to a seed supper over by Prairie Point about a week ago," Del remarked to one of the men. "Wasn't but a half dozen guys there. Not many over there doing row-crop anymore."

"Hard to make a dime row-cropping anywhere now," the other man said.

"A lot of them have given it up for cattle and fescue."

"Could be a good move. Maybe these guys here tonight are just harder-headed and don't know any better. Don't know when to quit."

"Well, land's better around here," Del continued. "Over at Prairie Point there's more brush and timber. Soil's not as good neither. Some guys say they can make more money selling land to city folks looking for little retirement places than farming. That's what some are doing if they own rough land."

"I don't doubt it. Hell, if you make any money at all, it'll beat farming," the other said. "I'm surprised any of us are still here doing this."

Seed suppers were an institution across the rural heartland of America, but Del figured that few people outside of agricultural communities would even be aware of them. They weren't much different than investment companies and real estate agents putting on free seminars with food. Seed companies held them every year about this time, but there was always a local sales rep present who was a farmer in the area. Everybody would know him. As the regional franchisee, he was the guy to contact when you needed seeds for spring planting. He planted test rows of a variety of seeds on his farm and marked the rows with little signs so it was easy to compare the different varieties. Everybody knew what the soil was like on the rep's farm, too, so they could make mental comparisons with their own farms. The seed suppers, however, were run by a regional company man who was likely not from the area but was from a rural background.

The company rep tonight was a young, heavy-set fellow from Nebraska in jeans and a checkered, western-cut shirt. He had a round face and a big smile, and he moved quickly through an array of seed products, making suggestions and asking for questions. There were plenty of questions.

Farmers are a suspicious lot, cautious in making decisions that could cost them money, and coy in dealing with large, impersonal corporations, which they seldom held in a favorable light. Seed suppers helped mitigate these suspicions because of the presence of a local rep, as well as the regional guy. The supper and informal discussions created a sense of community and camaraderie among the men, and the free meal didn't hurt. Everybody was in it together because they all had crops to plant.

Seed promotion had gone modern, and tonight computerized printouts were given to everyone to show yields on the local rep's test plots, as well as those on farms of a dozen other local reps in the bi-state area. There was data on maturity lengths—the consensus being 107 to 112 days for corn to develop at this latitude, 62 days to mid-bloom for milo, and so on. They also discussed seed dry-down times after maturity, disease resistance, tolerance to drought, and, of course, genetically engineered Roundup Ready seeds like soybeans.

Everyone in the room was familiar with Monsanto's Roundup Ready seeds, which were resistant to glyphosate herbicides. These were an enormous aid in the battle to control weeds. Most of the men in the room were planting them. On the other hand, genetically engineered seeds that protected against insect damage generated much more controversy. Many farmers were suspicious of these transgenic products, popularly known as Bt seeds, because the genes of a bacteria, embedded in each seed, turned all parts of the plant into biological pesticides. Concerns about the marketability of some of these bioengineered products were clearly on their minds. They'd heard about the disaster with StarLink, a Bt variety of genetically engineered corn seed. Some of it got accidentally mixed into corn tacos and food for human consumption, which led to it being pulled off the market. It could still be fed to animals and used in non-food industry, but nobody wanted to grow it now. Tonight the farmers were still grumbling about this incident. One farmer summed it up: "Won't do us no good to grow this stuff if nobody will buy it."

Another countered, "Well, right now it don't matter all that much 'cause the government still guarantees us a price for it."

Still another said, "Yeah, but the government ain't gonna keep sending us money forever if there's no market for this stuff. Eventually we're headed for a brick wall. Japanese don't want those genetically modified seeds, Europeans don't want 'em, neither, and if everybody's growin 'em, it gets harder to find seed that's not contaminated."

Pointing a finger at the two reps at the head of the room, somebody else quipped, "Even you seed companies can't guarantee your seeds are one hundred percent free of these modifications anymore. Ain't that right?"

The regional rep protested but agreed that it was getting harder to avoid contamination.

The farmer continued, "I hear you can get contamination from cross-pollination with a neighbor's crop. And also, contamination can happen if we don't get our planters and trucks all cleaned out good. Then the only thing we got left to sell is these genetically modified seeds. Then what are we gonna do?"

There was a moment of silence and some heads nodded. "He's right," someone said.

The revolution in seed genetics was clearly on the minds of many of the men at the supper, and not everyone was comfortable with the way things were headed, even if they used the seeds themselves. Most of the seed products offered that night were not the new genetically modified varieties, but a lot of guys could see change coming.

Seed orders were light, but that wasn't unusual. They would be ordering later because the deal was sweetened with a sixteen-percent discount for orders before the end of March. It was a good discount, especially with prices topping a hundred dollars a bushel for some seeds. The men and women finished their meal and coffee, and then it was all over but for the free caps with the company emblem on them that were handed out at the door. Everybody left with one or two caps, some of the wives claiming a cap as well. As they stepped outside, the hard frozen ground and the rush of cold air on a dark winter night made spring planting seem the stuff of fantasy, but Del knew better. In a few months he'd be placing his seed orders. He was already servicing his machinery, prepping it for the coming season. He'd be ready. They all would.

4

December 16
Maximilian

It was overcast and barely light when Zach pulled into the driveway, the tires on his brown feeding truck cutting lines in the snow from the second storm of the winter. Leaning out the open window on the driver's side, he called across the driveway to Del, who was getting into his feeding truck parked in the open-sided barn opposite the house.

"You ready to feed cattle?" he said, waiting to make sure Del's feeding truck would start.

"I'll take care of the cattle on the Somerset property and the Dunstan and Hyatt farms," Del responded, his voice clear in the cold morning air. "Can you get the rest?"

"No problem," Zach answered, hardly waiting for a response as he circled the driveway a little too fast, snow spewing from his tires.

Zach was twenty-nine, married a little over four years now. He'd met his wife, Amber, at a tractor-pulling event and joked that he wasn't sure if she'd fallen in love with him or his big red tractor. They lived two and a half miles away on the Hyatt farm, one of the properties they rented. Zach was stocky and strong, with arms like posts and a shy smile that came easy to his broad face. He was soft-spoken, loved farming, and possessed a natural aptitude for mechanics. He'd grown up farming with his dad and never really wanted to do anything else. He was also an accomplished welder and occasionally helped neighbors with welding jobs, but farm work didn't leave him much free time for outside jobs. When he did have extra time, he most often spent it working on his pulling tractor for competitive pulling contests. A few months ago he'd started building a new pulling tractor more powerful than anything he'd previously tackled.

Amber, on the other hand, had grown up in small towns and never lived in the country until she married Zach. She was surprised that she liked living on a farm. There had been doubt. With sparkling blue eyes and silky blond hair cropped pixie-like, she helped out around the farm when she could, but with a new baby and a job as a doctor's assistant in a nearby town there wasn't much free time.

Flipping her hair to the side, she laughed and told a former high school friend, "We won't get lonely out here. There's plenty of cows for neighbors."

Still, she worried when Zach forgot to call and tell her when he would be home. When they were first married, it had been difficult adjusting to Zach's long and often irregular hours. She had eventually reconciled her misgivings, realizing that schedules and regular working hours were impossible on a farm.

The feeding trucks—robust four-wheel drives with beefed-up springs and growling engines—were equipped with two hydraulic-powered steel arms at the rear of their flat steel beds. The arms were designed to clamp super-sized hay bales in a viselike grip and lift them upward onto the back

of the truck without an operator ever needing to touch a bale. There was just space enough for two bales, one forward and a second behind that was secured by the lift arms. To farm boys of earlier generations, it would have seemed a miracle—hauling a ton and a half of hay without breaking a sweat, all from the comfort of a truck cab with little more than a glance in the rearview mirror. The genius of Pascal and his principle of transmission of fluid-pressure would be helping to feed cows all across the Midwest this morning.

On a rental farm a mile from the homeplace, Zach backed up to a line of giant hay bales stored in the open. Snow blanketed the tops of the bales like icing on giant pastries. In quick succession, using the hydraulic lift arms, he hoisted up two bales. Back on the snowy road, he headed south to a T-intersection and turned left. A quarter mile beyond he opened a wire gate and drove into a pasture. The ground was frozen, and the truck bumped slowly across the snowy pasture, the snow disguising myriad frozen hoof prints that pockmarked the ground. Zach angled the truck toward the west side of the pasture where there was a tree-lined ravine. The pasture was strewn with cattle, motionless black lumps in blue-white snow. Standing in the frozen dawn, they watched the truck advancing across the snowy pasture. Then, almost as one, this lethargy of ungulate lumps began moving, slowly at first, then with a quickening pace. Soon they were streaming toward the truck, moving briskly, some even trotting, foggy breaths blowing in plumes. Within minutes they coalesced in noisy, bawling, dark lines behind the slow-moving vehicle, now an agrarian Pied Piper bearing a ton and a half of dry hay fit for a ruminant gut. Almost magically, twenty-five tons or more of bovine flesh had been mobilized, brought in tow, acquiescing to that most basic of animal needs—hunger.

Toward the western side of the pasture, Zach stopped the truck and got out, then walked a short distance and cleared a small area of snow away with his boot. He emptied a small portion of a fifty-pound sack of pellet supplement on the frozen ground and continued walking in a large semicircle, stopping several more times to scrape patches free of snow, and in each scrape he poured pellets. Responding to this weekly ritual, the cattle were as conditioned as Pavlov's famous salivating dogs. Almost immediately, it became a free-for-all, cows pushing, jostling like junkies in a frenzy of addiction to get at the bite-size mineral mix embedded in sweet-tasting molasses. Zach felt like he'd just given candy to a bunch of kids. By the time he finished emptying the sack, he could barely stay ahead of the melee.

Back in the truck, Zach drove ahead a short distance and released both bales near the shelter of the wooded ravine. He continued along the edge, checking to make sure a cow wasn't down in the ravine. It was too early for newborn calves, but checking stream banks and isolated corners of pastures was part of the job. It would continue throughout the winter calving season. He fed hay and supplement to two other herds, then ferried a couple of bales to six bulls on the homeplace. That was when he noticed a problem. During the winter, when the bulls weren't needed for breeding purposes, they were kept together in a pasture on the homeplace. But Del and Zach made one exception.

Maximillian, a venerable ten-year-old bull, stayed alone in a corral next to the barn, and that morning Zach discovered that Maximillian's problem had resurfaced again. He was a gentle giant, a huge Black Brangus with massive shoulders and a long, muscular body. Halter-broke to lead, he'd been shown at numerous fairs, even the American Royal in Kansas City. Now a few inches of Maximillian's soft, moist penis were exposed, and with the bitter cold of the last several nights, it was showing signs of frostbite again, the tip already turning slightly purplish blue. Zach winced just looking at it. Standing by the barn, head lowered, eyes rolled forward, Maximillian looked less regal this morning than his imposing size would have suggested. Zach opened the gate and herded him around the side of the barn and into the long narrow alleyway leading to a red squeeze chute at the front corner of the barn. Zach walked behind Maximillian, prodding him occasionally with a stick. The old bull lumbered forward, his damaged member swinging side to side. Immobilized in the steel squeeze chute, Maximillian knew the routine and remained calm, but Crissy, Del's over-exuberant border collie, heard the noise of the chute gate closing and rushed headlong toward the barn, barking furiously.

Zach quickly intervened, stopping her and yelling, "Crissy, get outta here! Quiet!" He grabbed her collar and tugged her off toward the shop to chain her to her doghouse. The last thing he wanted was a frightened bull that probably was strong enough to break the chute and maybe hurt himself doing so.

Zach was frustrated with the dog. Del had bought her as a pup, hoping she could be trained to help with cattle roundups, but she hadn't shown the behavioral traits trainers wanted to see before training border collies for herding. Her coat was glossy and black and set off by white feet and a white chest, and, like all border collies, her energy was boundless, but, in her case, it was largely unfocused. Zach muttered, "Bark when we don't want you to, but won't do a thing when we need you."

Zach prepared to deal with Maximillian's problem by bringing a small bucket of lukewarm water from the house. He pushed the little bucket into the bottom of the cattle chute and held it up underneath the bull and tried to submerge his frostbitten member in the water. Zach hoped it would warm the frostbitten tissue, but he wasn't sure if it really helped. Minutes later, Del arrived in his feeding truck, spotted Zach, and guessed the problem even before he reached the cattle chute.

"He never was much of a cold-weather bull," Zach said, looking up as Del approached.

"Yeah, this has happened before," Del said, grinning at the old bull's discomfort. "From here on we probably better keep him inside the barn. He does better that way. Won't be no good to us if his business freezes off."

Del moved to the edge of the squeeze chute. Kneeling beside Zach, he reached through the steel bars on the side of the chute and tried to push Maximillian's frost-damaged genitalia back up inside the sheath for warmth, but the indignity was too much for Maximillian, who struggled and let go a shower of urine that sprayed in all directions, dousing Del's hands and sleeves.

"Guess he don't like all this attention," Del shrugged, not showing much reaction to what had happened.

Zach grinned, then looked up. Through the open side of the barn, he saw the veterinary's pickup truck disappearing down the blacktop road beyond the house and over the hill, portable chute rattling and bouncing behind. "Hey! There goes the vet!" he exclaimed.

"See if you can get him on the phone," Del said. "Maybe you can stop him before he gets too far down the road."

Reaching inside his heavy coveralls to extract his cellphone, Zach fumbled with the keypad, cold fingers scrolling stiffly down the menu. By the time he got Dr. Reid Barkley on the phone, the vet was a mile down the road.

"Fellows, I've got a herd of calves to vaccinate and work this morning and I'm already late," he responded, his voice breaking up from the noise inside his truck. "I'll stop by this afternoon on my way home. See what I can do then."

Del nodded. "Nothin' more we can do here 'til he arrives. Vet's seen about everything. I reckon he can fix ol' Maximillian up again."

5

December 20
Fences

It was noon, and conversation around the kitchen table centered on a recent snowstorm and the cattle. More than anything, however, everyone was talking about the arrival of more new Amish neighbors.

"Five families moved here this year," Zach said. "That's on top of those already here. Two more coming in February, people are saying."

"I heard these new ones mostly came up on buses from Tennessee. Bought land as soon as they could. A few of 'em are from Minnesota too. They didn't quibble about the price of land, so you know they bought high," Del added.

"Do you know how they moved their things, them not driving trucks or cars or anything?" Caitlin asked. "Too far to go in a buggy."

"Shipped their farm equipment and farm animals by truck," Del answered. "At least all the heavy stuff like horse-drawn ploughs and planters and buggies came that way. Put their clothes and household things in big plastic fifty-five-gallon barrels. I seen a bunch of them blue barrels sitting out by one of their houses a few months ago."

"I guess compared to most of us they really don't have that much, anyway," Caitlin added.

"Prairie Point's finally putting up some hitching rails in town," Zach said. "I think they got five now. Haven't been any of those in town for almost a hundred years. I even saw one in front of the bank."

"County's been putting up those yellow horse and buggy signs too. You know, the ones that say 'Share the Road.' First time we've ever seen those," Caitlin said.

Harlan, who usually ate his midday meal at the farm with Del and Caitlin, spoke up as he was finishing his second bowl of Caitlin's chili. "You know why they're moving up here? The rough land they're buying east of town ain't all that great."

"Nobody seems to know for sure. I heard there was a dispute in Tennessee over those bright orange 'Slow Moving Vehicle' signs on their buggies," Caitlin said. "They don't like bright colors, but there was some state requirement. Never know how something like that is going to end up."

"I haven't seen any buggies around here with those orange signs," Zach said.

"I heard land prices were getting out of reach in Tennessee," Del added. "With their big families, I suspect they needed new land too. It's cheaper here even with them paying premium for it."

"They'll do all right. They work all kinds of jobs besides farming—train horses and mules, carpentry, finish cement, saw lumber. Some even do small engine repair. Women hire out to clean houses, all kinds of things," Zach said.

Harlan shifted in his chair. "Yeah, and they grow a lot of vegetables to sell too."

"And the women got their kids selling baked goods along the highway during the summer," Caitlin said. "Everybody does something. But they won't drive a car or truck or anything."

After the meal, Caitlin brought her grandson Nathaniel to the table. Chubby and with blondish hair, he was wide awake despite not having slept this morning.

"He slept ten hours straight last night," Zach said. "That's the first time he's done that in a while." Then he handed Nathaniel a little baby rattle, which the child promptly dropped on the floor.

"Here, this won't do you no good on the floor," Harlan said, picking the rattle up and handing it back to Nathaniel, who took it and dropped it again.

"Hey, this ain't no game," Harlan said. "Well, I was getting sleepy sitting here, anyway. Now, bending over to get your rattle twice has woke me up. Guess it's time to go do some work."

The men filed out of the kitchen and onto the back porch, which served as a sunroom for the cats, as well as mudroom and greenhouse. A line of heavy coats and coveralls, some nearly new, others worn and soiled, hung from pegs on the west wall. A metal International Harvester sign with a red tractor on it was tacked to the wall above the coveralls. The floor was littered with old boots. Several chairs, a rough-painted wooden bench, empty bushel baskets, a box of recently purchased fruits and vegetables, an array of potted plants, and a chest freezer competed for the remaining space. An antique wooden walking plough, nearly hidden by a large potted plant, leaned against a far corner next to the window. Toby, one of their cats, was asleep in the south-facing windowsill, his gray back stretched out to catch the sun's winter rays. As soon as he realized the men were looking at him, he jumped down and crouched under a chair in a dark corner, only the white on his face showing and his yellow eyes glaring at the men.

Del glanced at Toby. "He always was a goofy-acting cat; kinda wild sometimes. He don't like people much. Caitlin and I are about the only ones that can handle him. And he don't even like us all that much."

"He sure hates me," Zach said. "When you and Mom were on vacation last summer and I stayed over here part time to watch the place, he wouldn't come near me."

"He's spent almost his whole life here on this porch. He's a weird cat. Don't have no tail, neither. Not even a bone where his tail ought to be," Del added.

Nobody fretted much about Toby's lack of pedigree. He was fed and treated the same as every other animal around the place whether they spent their lives inside or out.

The men pulled on insulated coveralls, jackets, and heavy insulated rubber boots. "I hate wearing these rubber boots for making fence," Del said, struggling to get his foot in one of them. "Always seem to get a hole poked in them around barbed wire, but there's too much snow out not to wear them."

"Winter's a pain in the ass," Harlan added. "Always putting on clothes or takin 'em off. Sometimes you got so much on you can't hardly move or do anything. Taking a leak's an ordeal."

Zach laughed. "Harlan, it's not that bad. You just don't like cold weather."

"Don't I know it! Gloves are the worst. Can't do nothin' with gloves on. Fall down 'n' you gotta take 'em off first just to get up."

Outside, snow covered the ground. In a storage area attached to the farm shop, the men began loading bundles of steel fence posts, heavy spools of prickly barbed wire, and tools onto a flatbed pickup truck. Before they finished, a shiny black tractor-trailer with a sleeper and gleaming silver grain trailer pulled into the driveway.

"You selling more of your milo?" Harlan asked as the truck stopped beside one of the steel grain bins.

"Red milo's $5.10," Del said. "Better than harvest price, even after trucking, which is seventy cents a hundred. I'm gettin' rid of all of it."

"Spencer's gonna be sending trucks out every few days to load it out," Zach added. "Ought to get it cleaned out by the first of January."

Del walked over to greet the driver. "Where you headed with this load, Jim?"

"Everything's going to Memphis now. Don't know where it goes after that. Maybe down the Mississippi [and then] to some foreign country. It's an eight-hour run to Memphis." Jim was in his early sixties and still driving trucks full time. He wore gray, zip-up coveralls and leather Durango boots. A feed company cap shaded his slender face, which was creased with lines like pleats in an old curtain. His grin revealed a clutter of stained teeth that had held too many cigarettes, and there was a dark gap where a tooth should have been. But his quick humor and affable nature confounded first impressions.

"Then where you going?"

"I'll likely pick up a load of cottonseed meal in Memphis. Been taking a lot of it up to North Dakota. With luck bring a load of sunflower seed back down here somewhere. Try to get in two runs a week."

"Whee-oo!" Del whistled. "That'll keep you out of trouble."

Del figured a lot of his milo would end up as cattle feed. Maybe some would get exported overseas like Jim said, but it wasn't a big market. Cottonseed meal and cottonseed oil were bigger. They got used in a lot of animal food products like high-energy cattle supplements. Del sometimes fed a cottonseed supplement mix to his own cattle.

Del turned a mechanical crank to raise the auger. When he got it high enough on its stilt-like base to clear the side of the truck, Jim eased the truck backward so the auger pipe was centered over the long trailer. Looking like a giant mechanical insect, the auger rested atop an A-shaped steel frame that hinged on a two-wheeled base. An old red Farmall H tractor sitting next to the auger provided the power. Back by the steel grain bin, Del switched on an electric motor that powered a self-unloading auger inside the bin. Within seconds, a thick stream of milo gushed from the bottom of the big grain bin into the metal hopper of the auger. Then, without climbing up to the tractor seat, Del turned the key to start the little tractor. It chugged to life, its power shaft at the rear turning the auger's long Archimedes' screw. Moments later, a thick stream of reddish grain poured into the truck. Del and Jim stepped back out of the milo dust, which quickly formed a nearly opaque cloud that enveloped the truck and grain bins and began drifting northeastward toward the farmhouse.

"What happens if you overload your truck?" Del teased. "Will they make you scoop some of it off? You don't have no belly dump on that trailer."

"Hell no. I don't scoop nothin'," Jim said. "My part of the deal is to drive. Their part of the deal is to make sure the damn floorwalkers work right to unload the truck. If the load's too heavy, I can unload some at the elevator when I check the weight. You seen these self-unloading shuttle floors work, right?"

"I've seen 'em. Trailer floor moves up and down like a woolly worm crossing the road."

"Slickest damn thing you ever seen. It'll walk anything outta the trailer in no time. Walk you out the back, too, if you was in there. Unload this milo faster than most elevators can handle it. Except the big ones like at Kansas City. Those up there can take a whole load at once."

"Tell me when you think you're loaded," Del said, talking loudly over the noise of the auger.

"Let me check the pressure sensors. These damn things ain't perfect, but when the truck bed's level, they'll show if I'm near maximum weight."

Jim walked over to the truck cab and squinted at a couple of digital gauges. Sensors connected to the truck-axle air suspension system would give an approximate read of the load weight. "Don't want to get over the limit. Otherwise I'll be getting out my checkbook at the weight station," Jim said, grinning and showing the gap in his teeth.

The men stood to the side for another ten minutes and talked and waited. Then Jim checked his gauges again. "Hey, it's reading right at the limit. Shut 'er down. I think she's loaded."

Del walked quickly back to the steel bin and flipped the switch on the electric unloading motor that powered the auger. Standing by the rear wheel of the tractor, he waited a minute until the auger was free of grain before reaching up under the fuel tank and turning the tractor's ignition off. Abruptly, the tractor engine and clatter of the auger went quiet, and the cloud of milo dust slowly began to clear.

Jim was already turning a long-handled crank at the back of the trailer to unroll the trailer's tarp. He yanked the anchor ropes tight at the back corners of the trailer, securing the tarp so the load would be protected.

Watching him, Del said, "Once I didn't tighten down a tarp on a truck. That was back in the early days when I couldn't afford a fancy roll-up bar like you got. It was a lot of work to get up there and pull on that heavy tarp. I didn't have far to go, so I thought if I went slow it wouldn't blow off."

"Ha," Jim laughed. "Bet you learned."

"I sure did. Tarp was so heavy I thought I'd never get it back up on the truck. Zach was too little to help. That's when I got a roll-up bar. Wasn't gonna make that mistake again."

"Same thing happened to me once too," Jim said. "Tarp wasn't fastened good. Got loose before I noticed it. Dang thing was flappin' to beat hell. Tore almost plumb in two. I had to let it drag on the ground alongside the trailer until I got to the next town. That was some sight."

"Well, see you in a few days," Del said.

"Sure thing," Jim yelled, climbing into the cab.

Two thin columns of black smoke spiraled up from the rig's exhausts as it lurched forward, the cab bucking slightly. After leaving the driveway and turning eastward in a big arc, the truck slowly moved forward, then picked up speed. Del could hear Jim running up through the first of about a dozen gears as he topped the low ridge to the east. Forty-thousand pounds of red milo was quickly movin' east.

"Those guys are always in a hurry," Del said, turning to Zach and Harlan, who had finished loading the fencing equipment and wire and tools on the black flatbed pickup trucks. "Some drivers don't get paid a regular salary. They get maybe thirty percent of what the truck makes, so the more they're haulin', the more they're makin'. A lot of truck outfits, especially smaller ones like Spencer's that Jim works for, operate like that."

The snow was a great transformer, fashioning monochromatic art from trees and fences, almost anything that provided relief from pan-flat fields. It hid blemishes, softened rough edges, and painted gaunt, open spaces white, but its beauty was largely in the eyes of human beholders who could retreat from winter's extremes. To smaller inhabitants, hunter and hunted alike, it brought hardship. A northern harrier, a hawk elegant in silver, gray, and white, flashed black wingtips that looked as if they had been dipped in ink as it quartered low over the snowy terrain. Del barely noticed it as he drove westward along the blacktop road, which was still covered in patches of icy snow. His mind was on his afternoon's work.

The harrier hunted by flying low over fields. With its head bent sharply downward, its eyes roamed the terrain passing beneath, ceaselessly searching. Its eyes registered eight times as much detail as a human and missed little. With luck the harrier would surprise a careless rodent or unwary bird before sundown. Further down the road, a young red-tailed hawk sat stoically on a fence post. Its sit-and-wait hunting strategy contrasted sharply with the energetic searching and quartering of the harrier. The red-tail was larger and heavier than the harrier, not as agile, but capable of ambushing larger prey, and it would do this by flying out rapidly from a perch to pounce, or by soaring and then diving fast from considerable height. For now, all it could do was watch and wait. The snow cover made hunting small rodents difficult for both raptors because now rodents could forage out of sight beneath the snow, but the diverse mix of crop fields and prairie land on small farms provided each with favorable habitat for hunting.

By early afternoon, Del, Zach, and Harlan had two trucks, a tractor, and their fencing equipment in place on a rental property several miles west of the homeplace. Their aim was to finish a quarter mile of new fence before dark. On the short drive, Del shelled and ate roasted peanuts given out by the local co-op, and sipped a Diet Coke as he watched familiar snowy fields flow past his truck window. Del stashed the remainder on his truck dash and tossed the empty shells on the truck floor.

Earlier, Harlan had made fun of the shells on the floorboards, and Del had told him, "Harlan, you ain't seen nothing. Look in the vet's truck sometime. Last time I seen it, there was about two inches of peanut shells on the floorboards. Don't know how he drives with that mess in there."

The temperature hovered near freezing. For a while sunshine broke through the clouds, turning the snow-covered fields glittery white. Then clouds bunched up, leaden and lumpy, and a chill descended over the snowbound land. Two bulldozers, huge lumps covered in snow, sat mute at one end of the field. The smaller one, a bit of its red paint showing, belonged to Zach and Del, the larger yellow one to a contractor helping bulldoze brush and trees. They had reclaimed nearly ten acres of land this fall and planned to clear more. The new fence would mark the property boundary of the cleared land.

As they unloaded fencing equipment, Harlan asked about the guy that used to own the land.

"He never farmed much. Mostly worked as an electrician and television repairman," Del said. "Then he got married to a schoolteacher. Left the farm in such a rush he never even cleaned out the house."

"Probably wasn't worth much anyway," Harlan interrupted.

"The place grew up in weeds and brush and ever'thing. Thieves stole some stuff, but they never bothered the house none. Nowadays they'd burn down the house just for meanness. New owner wants to get it all cleaned up."

"That why he's spending all this money?" Harlan asked.

"He's spending a lot on bulldozing brush and having new fences built. But it's tough keeping fencerows free of brush. They don't stay clean without fire like what used to burn across the prairies and kill back saplings and brush. Now about all you can do is spray with brush killer unless you got horses or mules. They'll eat sprouts."

Most of the trees they'd bulldozed in the old fence line were hedge trees, also known as hedge apples. Botanists know them as Osage orange, *Maclura pomifera*. It has greenish fruits as big as grapefruits, a hard and rough-textured outside, and with lots of sticky white sap inside. Many of the trees were large—remnants of an era when farmers planted them for living fences and for wooden fence posts. Before steel posts became economical, farmers built a lot of fences with Osage orange. The wood was hard and durable but the twigs had glove-ripping thorns and the branches were crooked. The posts would last a half-century or more.

Fresh-cut Osage orange wood is beautiful. It shows a fine interior grain of rich, dark orange. Settlers extracted a yellow dye from the bark of the roots, but the trees were never used commercially for furniture because the wood always split when it dried. Worse, still, it is so hard and burns so hot in stoves it was a fire danger. On the other hand, the saplings and green wood are springy and resilient, attributes prized by Native Americans for making bows.

The original range of the tree was thought to be eastern Texas and Oklahoma and maybe part of Arkansas, but farmers planted them all over the Midwest, then took them east. They soon become a nuisance, spreading over pastures and ruining them for grazing. Now these old Osage orange fencerows were being bulldozed to reclaim the land, a last reminder of an era when manual labor was cheap and plentiful. Although Del, like almost everyone else, used steel posts for fencing, he still used Osage orange posts at the ends of fence lines because the big posts provided better support.

Using a chainsaw, Zach and Harlan cut six thick Osage orange trunks to use as end-posts, three at each end of the quarter-mile fence. The posts were huge—more than eight feet long, sixteen inches in diameter, and weighing several hundred pounds each. The men chained the posts to a hydraulic front loader on their tractor to hoist them into holes. Harlan and Zach worked quickly in the cold, shoveling dirt around the posts to hold them upright. Harlan cut a V-shaped notch on the inside of the two outer posts with a chainsaw, and a similar V notch about halfway up on both sides of the middle post. Each cut exposed some of the bright-yellow interior wood of the posts. He then fitted a slender wooden brace post between each pair of upright posts, sliding them neatly into the V-slots so they were held in place. Then he cut a second, narrower set of grooves in the posts, this time on the outside of the outer post but below the middle, and a narrow groove on the far side of the center post above the middle. Then, wrapping heavy gauge wire diagonally around this pair of posts and sliding it into these newly cut slots, Harlan and Zach used short sticks to simultaneously twist and tighten the strands of heavy looped wire. They then repeated this

procedure with the middle and inner posts. This locked the three end-posts together with a pair of X-shaped braces between the posts that would hold fence wires tight for years.

Fence building, like many farm skills, is individualistic. Del, Zach, and Harlan, aided by a powerful posthole digger and hydraulics, erred on the side of overbuilding. When Del saw poorly made fences, he knew there was a good chance the farming was done the same way. He said it didn't show pride.

Overhead two flocks of snow geese, white with black wingtips against a gray-white sky, passed by in V formations with the larger V almost enclosing the second. The birds called steadily with monosyllabic *wonk* and *hank* notes as they rowed southward. Their flight was steady and powerful with individuals occasionally shifting positions in the V. It was a majesty difficult for today's urban dwellers to appreciate, who mainly knew flocks of geese through their now urbanized, free-wheeling cousin, the Canada goose. Mention the Canada's name and images of jaywalking geese blocking traffic and geese from hell descending en masse and crapping on golf courses and sidewalks would surely come to mind. Moments later, another group of snow geese passed, then another. The birds had an innate ability to sense changes in the weather, something the men on the ground could not know without modern weather forecasting. Del was, however, aware of an approaching change in the weather. He and Harlan paused only twice to drink cold Diet Cokes during the afternoon. They knew the ground would freeze hard that night, cutting short their fence-making efforts, so they worked feverishly through the afternoon and into the gathering darkness, rarely pausing to look up, even as more flocks of geese passed overhead.

At the back of the flatbed pickup truck, Del suspended a prickly spool of shiny barbed wire by a steel bar that allowed the spool to rotate freely. With a strand of wire tied around the endmost fence post, Del drove slowly along the projected fence line, the spool unrolling behind, clicking and clacking and occasionally jerking when razor-sharp barbs snagged, causing the wire to catch momentarily. In a few minutes, he unrolled the quarter-mile length of wire, then threaded the end of the wire into a "come-along" attached to the outermost support post and tightened the wire. The come-along, a ratchet-like device similar to those used on trucks to tighten straps around heavy loads, quickly pulled the wire tight. Now the three men had a straight line guide to set the steel posts.

At the far end of the fence, one end-post remained to be set. Del wanted to get this last post in the ground so Zach and Harlan could add the braces. Something, however, was preventing the digger, a heavy Archimedes' screw almost two feet across, from going as deep as needed. The hydraulic-powered digger was one that Del and Zach had built. Suddenly it began to vibrate violently, then struck a tree root, and a half-moon keeper linking the shaft of the hydraulic pump to a sprocket wheel broke. To replace the half-moon keeper, Del had to take the chain off the sprocket, a tricky maneuver in freezing temperatures. With fingers stiff and clumsy from the cold, he accidentally dropped a link of chain, and it fell into the pulverized earth at the bottom of the hole. He couldn't see the link in the fading light, and it was too far down to reach. He had an extra chain link and half-moon keeper in his toolbox, but the keeper wasn't the right size.

Disgusted, he said to Zach, "We'll have to take the digger back to the shop and fix it. Can't do nothin' more with it here. We need to make some changes to it anyway."

They'd gotten the digger from a telephone company, but the hydraulic pump and transfer case that supplied power to the digger shaft were homemade.

"It's still sort of a work in progress," Zach agreed.

"Forget it for now. Let's get these steel posts in the ground before dark."

Harlan walked ahead, plodding through the snow. He stepped off a distance of about two post lengths, then dropped each post into the snow at the appropriate point, green steel almost disappearing in white snow. Every few minutes he walked back, drove the truck forward, and got a new armload of steel posts. Following him, Zach picked up the posts, holding each upright and perfectly straight while Del positioned the tractor so its hydraulic front loader rested squarely atop each post. Then, lowering the loader bucket slowly, he carefully pressed each post down into the cold ground. In less than an hour they set the entire quarter mile, a picket fence of green steel, an arrow-straight line disappearing into semidarkness. By the time they finished, the temperature was below freezing.

A barred owl called from a nearby woodlot. Its long, laughing call, almost maniacal in the shadowy dusk, lent uneasiness to the cold landscape, but it went unnoticed by the men. It was late and they were cold and tired, intent now on retrieving their tools, unused fence posts, and keys to the tractor, which they left in the field.

"Good thing we got finished setting the posts," Del said to Zach. "Ground's gonna freeze hard tonight. Never get another post set tomorrow."

"Make sure we get everything picked up," Zach said. "It's almost dark."

"We used to never have to worry about that sort of thing," Del said. "We left stuff in fields all the time. Nobody bothered it. Couple months ago the Gannons left a post-hole digger in a field and somebody stole it. Nice Danuser digger too."

Zach was impressed that someone would attempt such a heist. "I heard two or three guys had to be in on it 'cause it was so heavy. There weren't any truck tracks showing. Those guys musta carried it a quarter mile."

"The Gannons said they called the manufacturer and every auctioneer and sales outlet around. Put out the serial number in case somebody tried to sell it. Nothing's turned up," Del added.

"Maybe the thieves are using it digging holes!"

Del laughed. "Huh! I doubt that. I bet somebody thought they could sell it."

On the way home, the heater quit in Del's truck. Cold air whistled in around the door and under the dash. At the shop he nosed the truck in between partly opened double doors and checked for current in the heater wires, quickly eliminating wiring and fuses as problems. "Heater motor's burned out," he said, somewhat disgusted. "No time to fix it tonight, but I got a salvaged motor. Maybe I can fix it tomorrow morning before I feed cattle. Don't wanna sit in a freezing truck cab all morning."

Zach looked at his watch. "I gotta go. It's been dark for an hour. Amber's gonna be wondering why we're so late."

"Me too," Del replied with some resignation. "Caitlin will be all over me if I don't get in the house and get cleaned up. She wants to do some Christmas shopping tonight. Been planning this night in Grand Forks for several days."

Zach called Amber as he walked back to his truck. After four years of marriage, Amber still wasn't used to his long and irregular hours. As Del headed to the house, he watched Zach's pickup taillights receding down the highway, red embers disappearing into the cold blue night. He guessed Zach was still on the phone.

6

January 9
Help or Handout?

January. Feeding cattle dominates winter work; farmers attend machinery and farm sales; overhaul and maintenance of farm machinery continues in farm shop; farmers meet with Farm Service Agency representatives to review federal farm support programs; soils tested and lime spread on fields; farmers continue selling stored grain if prices hold; first spring calves born at end of month; white-tailed deer bucks begin losing antlers.

A wind blew cold and hard across the brittle farmland. Patches of snow, remnants of a recent snowstorm, lingered in ditch banks and sheltered areas and in cornfields where the stubble showed in long lines. Houses and barns were buttoned tight. Del was in a hurry. He had a meeting with the county Farm Service Agency to enroll for crop supports, but cattle chores had taken longer than usual. At the homeplace, he was out of the truck cab and walking quickly across the driveway even before the truck rolled to a full stop. His insulated coveralls rustled as he headed for the house to shed work clothes and heavy boots. Boy the shop cat spotted him, came bounding along, and deftly slipped past the aluminum porch door before it closed.

Farmers across the country had to queue at FSA offices the first quarter of each year to review farm assistance program options if they wanted to participate in government programs. It was a winter rite of passage. A new farm bill meant more bureaucracy, more paperwork, and another trip to the FSA office. A few farmers eschewed assistance, refusing to participate. Others grew crops such as cucumbers or sunflowers or fruits and vegetables that weren't supported by government price supports. Most farmers that grew supported crops, however, figured the programs were worth the trouble for the payouts when prices were low or crop yields down. It was supposed to be a safety net for farmers, but the program, a holdover from the 1930s Depression era, had changed dramatically. Now corporate agribusiness had gotten involved, as well as corporations and wealthy individuals who weren't even in agriculture but were large landholders.

Del was ambivalent about farm support programs in their present form, but he figured that if the government was going to give out money he'd get in line. On the way to his appointment, watching the frozen landscape flow past his truck window, he had a few moments to reflect on farm programs, and to prepare mentally for the meeting. Farm support programs are more complex and burdensome now than in the early years of the program, and more often than not they benefit corporate agribusiness more than the small farmer who sometimes actually needs them. He'd heard about the latest round of changes and figured he was about to see, once again, just how complex the program had become.

Farm support payments had been part of Franklin D. Roosevelt's New Deal efforts implemented during the Great Depression to reduce farm overproduction and to shore up collapsed commodity prices. Under Roosevelt's plan, the government set a target price for a commodity, and if the market fell below that target, the farmer didn't have to sell at below market price. Instead, he had the option of taking out a loan for the full target price value of the crop. The farmer then stored his crop as collateral against the loan, and held it until the market improved. If it did, and the market price exceeded the government's target price, he could sell the crop, which would pay off the loan, and he could keep the additional above-the-target-price money. If the market didn't improve by the end of the year, the farmer gave the crop to the government and kept the loan as payment in full for his crop. The government, for its part, stored the crop—now officially part of a grain reserve managed by the Department of Agriculture—until such time as market prices improved. By judicious selling from these reserves, spikes in commodity prices were moderated and food costs held relatively steady. Sometimes the federal government even made money on the commodity sales. The overall cost to the government for the program was, in present-day dollars, a fraction of today's cost for farm supports. The program also had a stabilizing influence on commodity markets. This helped farmers and American agriculture, but it was unpopular with speculators who often realized large profits from wild swings in commodity prices.

All of this was dismantled in the early 1970s when a combination of huge grain export sales and several drought years put upward pressure on commodity prices. Consumer food prices spiraled upward, generating political heat and consumer protests. To deal with these problems, Earl Butz, then secretary of agriculture, issued his famous proclamation urging farmers to plant their fields "fencerow to fencerow." Butz scuttled the government granary, lowered target prices, and put in place a subsidy system that gradually replaced loans to farmers with direct payments to them. In essence, the government wrote checks to farmers who, in turn, dumped massive amounts of grain on the market regardless of the price. His efforts dismantled a system that, although imperfect, had for forty years largely prevented food overproduction. The changes implemented by Butz didn't seem that dramatic at the time, but they worked almost too well. Farmers were swept into a downward spiral of commodity prices beginning in the late 1970s that continued into the early part of the twenty-first century. They increased production just to stay even. This led to continued overproduction and kept commodity prices low. In recent years, however, unstable geopolitical events, programs to convert corn into ethanol, and soaring world food demand combined to exert upward pressure on commodity prices once again.

Butz's policies resulted in a shift from supporting commodity prices to subsidizing them as they trended lower. It was an enormous boon to agribusinesses who could now buy raw commodities at bargain prices, and agribusiness lobbied hard for farm policies that kept commodity prices low. Now farmers were increasing production and driving down prices. Instead of supporting farmers, the government was supporting corn production. Overproducing corn made sweet food cheap, one of the reasons high fructose corn syrup was so widely used. The cheap-food farm policy, however, came with a steep price tag, some 190 billion dollars a year in federal farm price supports. And there was an even darker side Del had read about. Low corn prices made nutritionally poor food containing sugary products cheap compared to nutritionally high-quality food such as fruits and vegetables. And it was the nutritionally poor foods, typically high in sugars, that had been implicated in an increase in obesity. It almost looked like farm policy was subsidizing an industry that was linked to the growing obesity problem.

Del had seen the figures. The largest farms, the top ten percent, which included many corporate-sized agribusinesses, received almost eighty percent of all federal farm subsidies. Small farmers got relatively little. Federal subsidies were paid on only about fifteen farm commodities, and ninety percent of the subsidies were paid on just five crops. Of those five, Del grew three: corn, wheat, and soybeans. The other two, rice and cotton, were southern crops. Most of the remaining 400 or more farm products produced around the country received no federal subsidies. It was an irony, perhaps, that subsidies seemed to lead to overproduction and lower prices, while the unsubsidized portions of American agriculture—e.g., livestock, poultry, most fruits and vegetables, oil seeds, grass seeds, and so on—had, overall, done better, price-wise, than those receiving subsidies. The federal farm program needed an overhaul. Yet for all its faults, Del thought it would be shortsighted, in the near-term, to scuttle the programs completely. That could put a lot of the farm industry at risk.

Critics of farm supports also misunderstood the legacy of farming as a craft. Family farms were more than a business. They were the culmination of a life's work, often generations of work. Farmers almost always came from a family heritage of farming. They were born into it and began learning their skills at an early age. For those that stayed in farming, success often came slowly, sometimes only with land being passed from generation to generation within families. Capital outlays needed to succeed in farming were prohibitively high, and the knowledge and experience needed to succeed were not easily or quickly learned. There were no trade schools where you went to learn how to be a farmer. In farming, one was either in it for life or out of it forever. One could not move in and out of farming with the shifting winds of government policy.

Farmers also resented being used as bargaining chips in international policy. It had happened in the past when, for example, Jimmy Carter used a grain embargo as a weapon against Russia. Del remembered that fiasco all too well. For that reason alone, he figured, agricultural subsidies, odious as they were to critics, probably ought to continue to play a role, although perhaps in a form more similar to pre-1970s policies, meaning as a safety net. The alternative, some farmers felt, would be abdication to corporate agribusiness, loss of the American family farm, and with it more erosion of the American public's control over food pricing.

In 2006 Washington mandated the addition of billions of gallons of corn-derived ethanol to gasoline, and after that corn prices began to rise. This, combined with increasing numbers of middle-class consumers in emerging economies, put upward pressure on corn and some other commodity prices. It was a windfall for farmers, but it had been a long time coming. For decades, grain farmers had depended upon relatively low government subsidies for up to half of their income, even as operating costs soared. Now, thanks to increasing commodity prices, government support subsidies were well below market prices, and farm subsidy payouts in the six-year period from 2005 to 2011 declined by fifty percent, and could shrink by half again. This, as Del knew, was good news, but talk of scuttling farm subsidies entirely made him nervous. Farm commodity prices were notoriously volatile, and Del, like most farmers, was skeptical that the high prices would last, in part because it would only encourage others around the world to ramp up production. American farmers didn't need the subsidies now, he reasoned, but they might in the future.

The nation could continue without small family farms, but its people might not like that future. Small family farms brought diversity and competition to markets, while the increasing concentration of farmland and farm products into fewer and fewer corporate entities did the opposite. It reduced competition and placed increased reliance for a basic need—food—into the hands of fewer people. Once lost, the balance and diversity that small family farms brought to American agriculture would be difficult to replace. It provided valuable checks and balances against the increasing control of food by a few mega-corporations. The question of who raises food and processes and distributes it, as well as who owns the farmland and supplies the materials that go into raising that food, were questions that ought to be on the mind of every critic advocating abolition of farm subsidies. That's what Del thought, anyway.

The risk of being held hostage to food, and its distribution, was a question that Del thought should concern everyone. Were people willing to sacrifice the diversity of smaller farms for a few dollars in savings at the supermarket counter? That, of course, was where government farm policy entered the equation. If there was going to be hope for the long-term survival of small and mid-size American farms, there had to be even-handed, long-term agricultural policy that would benefit or at least level the playing field for the smaller farmers.

Some bureaucratic planners dismissed the value of small family farms, saying if they weren't efficient they didn't deserve to survive. While not all small farming operations are efficient, moderate-sized family farms can be as efficient as corporate operations. In some cases, the pendulum had already swung in favor of middle-sized farm operations. Studies showed that the cost of production of some agricultural products was now lowest on moderate-sized, family-run farms up to about 2500 acres. Where corporate businesses gained, however, was in marketing, which continued to increase with size. It was more efficient for wholesalers to purchase from one or a few large corporate operations than from many small farms, even if they purchased at a higher price. Del wondered what price future generations would put on this loss of diversity.

The Department of Agriculture was reorganized in 1994, and the Farm Service Agency (FSA) emerged, incorporating an alphabet soup of earlier agencies including ones that administered Roosevelt's original programs in the 1930s. From the farmer's perspective, the programs hadn't really changed much since the reorganization. Whether it was land set-asides, crop deficiency payments, disaster relief, supplemental revenue assistance, direct and counter-cyclical programs, or any of several others, they were mostly just more complicated now. Del, like many farmers, also felt the bill was misunderstood by the American public because recent farm bills weren't just about farming. They included money for food stamps, energy, telecommunications, and forestry and conservation, among other things, and Del thought farmers were getting a bad rap for an expensive government program that had less to do with farming than it did with a lot of other things. Food stamps were a particularly sore point. Del had read that food stamp costs doubled from 2001 to 2006 and had continued to increase. In the 1970s, only about one in fifty people received food stamps. Today it was around one in seven, and half the recipients were staying on the program for eight years or more. He wondered what had become of the simple idea of a farm-only bill, one not hostage to so many other government programs.

The Middleton County FSA building was located a mile out of Bellamy, the county seat. It was a modern, single-story, red-brick building, half devoted to FSA business, the other half to county soil and water conservation. Del entered the FSA side. Even before he reached the long front desk, Lindsay, one of the county agents, motioned him back. She and several coworkers shared a large open room with workspaces partly separated by rows of file cabinets.

Lindsay's desk was piled high with notebooks and file-folders, but they were neatly labeled and well organized considering the number of files she handled. A fingerprint-smudged computer monitor on her desk was positioned so both she and the farmers she helped could view the screen. Lindsay stood up and greeted Del with a firm handshake and a smile. Having grown up on a farm in rural Missouri, she understood farming and had a lot of empathy for farmers and the issues they faced. She also genuinely seemed to enjoy helping them navigate the bureaucracy of compliance.

As Del sat down, he glanced at a piece of paper, dog-eared from handling by rough-calloused hands, taped to the side of a file cabinet by Lindsay's desk. He'd seen it before on the Internet and maybe printed somewhere; it was the kind of humor that farmers appreciated. It read:

Farmer's Last Will

1. To my wife I leave my overdraft at the bank. Maybe she can explain it.
2. To my banker I leave my soul. He has the mortgage on it anyway.
3. To my neighbor I leave my clown suit. He'll need it if he continues to farm as he has in the past.
4. To the FSA I leave my grain bin. I was planning to let them take it next year anyway.
5. To my county agent 50 bushels of corn. Maybe he can hit the market; I never could.
6. To the junkman, all my machinery. He's had his eyes on it for years.
7. To my undertaker, a special request. I want six implement and fertilizer dealers to carry my casket. They're used to carrying me anyway.
8. And to the gravedigger, don't bother. The hole I'm in is already big enough.

Del turned to Lindsay. "Remember a few years ago when some ol' boy got a list off the Internet of the federal crop payouts for farmers in the county? Had everybody's name on it, and he posted it in the feed store over at Ashton."

"I heard about that. Didn't surprise me none. It's public information if somebody wants to dig it up. Nowadays hardly anybody is getting crop payouts, what with the high prices and all, so nobody cares much."

"Everybody was stopping by to see where they were on the list and who was dipping the deepest. Some guys were pretty sore when they found out their government payouts were on public display."

"Oh, I know. You farmers tend to be pretty independent-minded," Lindsay said with a laugh. "I know you don't like your neighbors knowing too much about your business. Course, everybody can pretty much tell anyway. Now, getting down to business, this new farm bill is a lot more complicated than in the past. Doesn't surprise you, does it?"

"I heard it was the most confusing one yet."

"Gets worse every year. I'll try to explain it all," she said, putting a little empathy in her voice, "but I'm having to do a lot of handholding with farmers coming in here that don't really understand it. This year you've got five different options for computing acreage on six different crops—wheat, corn, sorghum, soybeans, oats, and barley. That's just for this region."

"I heard payouts vary from one place to another on the same crops."

"Yes, it's true. Can you imagine the government trying to track all that stuff?"

"Who knows what they do with it."

"Truth is I couldn't figure out any of these options without a computer. But I've gotten pretty good at ferreting out the ones that offer you the highest payouts for the least amount of paperwork. Some require you to prove yields for each crop. Means you got to keep track of every weight ticket from the elevator for each crop on each farm. I don't think you want to do that, do you? If I was you, I'd steer clear of that sort of thing."

Del frowned. "Sounds right, but I have to keep track of most of it, anyway, for all the rental properties."

Lindsay continued, "You remember that under earlier programs, once you signed up, if you wanted to be eligible for government assistance, you couldn't plant any more acreage than allotted under the program. It was figured on a base acreage determined by your past crop history for wheat and feed grains. The more you planted in the past, the more you were eligible to collect on now. At least the new farm program no longer links payouts to the amount of acres under cultivation."

"That was a real pain," Del said. "Often we had to split up fields. Otherwise, if we planted the whole field we'd exceed our allotted acres and lose our eligibility. Then the rest of the field would be in grass or something, and you'd have to mow it. It was awful trying to keep it all straight. Leastways now we can plant what we want. Just can't get subsidized but for so much."

Lindsay began sifting through the paperwork. "This is going to take a while," she said, pushing a curl of brown hair away from her face. "We have to do these one farm at a time. You want a cup of coffee or something before we start?"

"No, Diet Coke's fine."

"You're operating nine farms in addition to your own properties, so I'm going to be shoving a lot of papers in front of you to sign. Here's your acreage basis, as before, on the Somerset property. We'll start with that one."

The number of options was confusing, but once Lindsay went through a couple of the farms with Del the rest were similar.

Three hours and two Diet Cokes later, she and Del finished the paperwork for his farm and for the properties that he rented. He signed most of the documents under a power of attorney agreement. For three farms he'd have to send the paperwork to the landowners to sign.

At the midday dinner, Del sat around the kitchen table with Caitlin, Harlan, and Zach. Nathaniel was in a baby highchair. Amber usually left Nathaniel at day care, but when Caitlin wasn't working she often volunteered to look after him. Nathaniel took his baby-sitting options in stride. Exploration was on his mind.

"Time to baby-proof the house," Caitlin said, casting an eye at the refrigerator door covered with little red tractor and bulldozer magnets, and at the kitchen cabinets, which could be opened easily.

The family room was a bigger challenge—Zach's toy tractor and machinery collection from childhood still occupied a couple low shelves on a bookcase. There also were a dozen shiny, tractor-pull trophies, family photos, farm magazines, potted plants, and family mementos, all fodder for a toddler's fingers. Even Candy, the long-haired, black and white cat who often slept on a chair in the family room, was feeling insecure. Nathaniel couldn't resist pulling her tail.

After lunch, Harlan asked Zach to help him align a new power takeoff on Del's fertilizer truck chassis. With the two of them occupied, Del went to Ashton to buy oil filters and truck and tractor parts. Driving westward, watching the flat, brown countryside roll by, Del noticed a number of hawks on fence posts and in trees by the road. It was like that every winter. He thought many of them looked like young, naïve birds probably hunting too close to roads. Some would get hit by traffic before winter was over. Others wouldn't find enough food, or they'd get pushed out of good hunting areas by experienced birds. Come spring there wouldn't be so many, and there would still be plenty of rats and mice in the fields.

Del was back in the shop by late afternoon. With Zach's help, Harlan had the power takeoff installed, but now Zach was struggling with a transmission alignment problem on the pulling tractor he was building. Del also had a message from Caitlin saying she'd gone to a city beautification meeting in Prairie Point.

"Guess I'll warm up some stew tonight," he said to Zach. "She never could say no if there was somebody needing help."

Changing the subject, he said, "You want to go to the cattle auction tonight? Caitlin said she'd go if she got back in time."

"Can't. Amber's got something organized with friends."

"Well, I'm going for sure. Auction don't start until six. They usually don't sell anything important for the first hour or so anyway. Never know what might turn up for sale. Caitlin always likes to go too."

Like many farmers, Del and Caitlin often went to cattle sales even when they didn't intend to buy cattle. They did it partly to keep informed about cattle prices and the quality of cattle coming to market, and it also provided an evening diversion away from winter routines.

7

January 31
Winter Work

A farmer's world shrinks in the winter. Like mice burrowing under snow, a lot of winter farm work takes place out of sight. At this time of year, when fields are mattes of pewter and trees calligraphic etchings, farmers move mainly in lines between houses and barns, and to pastures speckled with dark, slow cattle. From a highway, travelers stare and see a bare, slightly undulating landscape shuttling past their windows with monotonous repetition. But they miss the lifeblood of the farms they pass because winter, more than any other season, is a time when much of the work on a farm goes unseen.

Beneath the masquerade of stark fields, brown pastures, and button-tight buildings, Del and Zach, as well as Caitlin and Harlan, continued daily farm chores even as the land around them acquiesced to winter's impoverishing hand. By now Del's winter farm chores were mostly routines that consisted of feeding and caring for his cattle, but the unexpected was always lurking nearby. Winter is stressful for domesticated animals, for humans, and for equipment. This winter three snowstorms had struck already. The night before was warm, and almost a half-inch of rain fell. Today a sodden mantle of gray clouds lay heavy over the farmland. Roads were wet and fields were glazed with a veneer of icy mud over frozen ground. Patches of ground fog enveloped low areas like cold, damp blankets. Even the sun breaking through the fog seemed suspended unnaturally low, an anemic yellow orb scarcely illuminating the land.

Del and Caitlin discussed the day's plans over breakfast. Their daughter Wendy was pregnant, and her baby due any day. Caitlin was excited, but Del sensed anxiety in her voice.

"Don't you and Zach go off somewhere too far away until the baby comes," she admonished. "I heard you talking about going off to go look for machinery parts at salvage yards. Some of those are pretty far away."

"We're not planning to do that for a few days. Maybe she'll have her baby by them."

"Well, what if she doesn't, and then you leave and it comes?"

"Oh, I know you'll call us," Del replied, not entirely sure he'd said the right thing.

"I want you around here," Caitlin responded firmly. "You know, like if something happens."

"Hon, I know you're nervous with the baby being so close, but we can't really do much about it. Wendy will be fine. She's having it in the same hospital where she works." Del's pragmatism was derived, in no small part, from having been so close to hundreds of births each year on the farm. He played a midwifing role himself to a lot of newborn calves each winter and spring, and he viewed the eminent arrival of this grandchild with similar optimism.

Caitlin knew that he cared, but she couldn't help being nervous. "I guess I won't completely relax until the baby comes," she replied.

"Me, neither, I suppose," Del added, trying to sound supportive. "I've got to renew my herbicide license at the county ag office this morning. That's not very far away."

"Just make sure you take your cellphone," Caitlin said as he left the house.

Meeting briefly by the farm shop, Del and Zach, both in heavy brown coveralls, hooded pullovers, and celery-green insulated boots, went over work plans as they did most mornings. "I got to take that exam for my herbicide license," Del told Zach. "Course, from the looks of the place this morning you wouldn't know I was ever going to need it."

"Ask them if I need one," Zach said.

"Don't think you will. Just one of us needs a license to buy it."

"Fine with me. No use in me spending another twenty-five bucks then."

"They'll make us watch a film. Then take the safety test. Almost every herbicide or pesticide purchase requires a license now. Can't buy it otherwise."

Satisfied that he wouldn't have to take the exam, Zach was already losing interest. "Pick up some creep-feeder supplement for the calves and a sack of dog food on your way back if you can."

"That little dog of yours sure eats a lot."

"She's a pup. Still growing."

"I wonder if she'll amount to anything when she grows up. Too small to be a stock dog."

"So far all she can do is bark and eat," Zach said, laughing.

With Del renewing his herbicide license, Zach took over morning cattle-feeding chores. Before leaving, he backed his brown feeding truck up against a recently repaired circular bale holder in front of the farm shop. It was just big enough to hold one large hay bale, but the holders got rough treatment from hungry, thousand-pound cows pushing and shoving. Sometimes they bent or sprung the steel frames. Using the hydraulic lift arms at the back of the truck, Zach hoisted up the holder and left the driveway, which was still slushy from the previous night's rain, and turned west onto the wet blacktop. With the holder suspended high behind the truck, it looked a little like a large communication device moving down the highway, but the only communication going on would be between hungry cows and hay bales.

All the cattle herds were checked and fed each morning and sometimes rechecked again in the evening, especially now that calving season was close. At first just one or two newborn calves would arrive each day, then there would be a half dozen or more. Busiest days were often the coldest or snowiest. When that happened, most of the day might be spent with the herds—checking newborn calves, temporarily taking custody of some that were weak and putting them in a barn or the farm shop to warm up and dry if they were cold or wet. They'd make sure their mothers accepted them and had milk, and that the calves were strong enough to suckle. They'd doctor them with pills if they had scouring diarrhea. Aggressive cows, overprotective of their offspring, often made things dangerous. Even gentle cows could be unpredictable when they had baby calves. By the end of calving season, there were usually several orphan calves. Del kept them in the barn nearest to the house, and Caitlin patiently bottle-fed them each morning and evening until they were old enough to fend for themselves. She said bottle-feeding calves wasn't her favorite chore, but watching them learn to feed was comical. The calves were always hungry and eager, but naïve and apt to spill or waste more supplement than they drank, so she had to watch them closely. Despite the work, she and Del figured that saving a few calves that would otherwise die was worth the effort. It made a difference to the bottom line at the end of the year.

By mid-morning, the ground had thawed on top, and Zach's truck fishtailed in the slippery mud as he ferried the last of nine giant bales of fescue, some thirteen thousand pounds of hay, to their cattle herds. At each farm, he checked water tanks and calf mineral supplement at calf feeders. Sitting in his mud-splattered truck at the edge of a broad, sloping pasture, he looked at the last herd of the morning. The only remaining chore was filling a feed trough with ground grain sorghum and supplement for the bred heifers at the homeplace. So far this morning, there had been no problems—no new calves, no major sickness or emergency, no cattle breaking through fences—and feed and water supplies were normal. One water tank float valve was leaking, and a partly frozen lagoon of water had formed overnight around the concrete watering tank. He'd fix that later. Then he noticed a cow limping, a black one with a white face. The limp wasn't obvious, a slight hesitancy in the cow's right rear leg. An unpracticed eye wouldn't have picked it out among the sixty cows in the herd. Staying in the truck, he drove closer, not wanting to frighten the cow. There was no swelling that might indicate a sprain. He figured it was bovine foot rot, a problem that occurred most often in the winter when cows walked on hard frozen ground and bruised or cracked the skin between their hooves. Then an infection started. The lameness made it hard for a cow to get around or get enough to eat. It was easy to treat if caught early. He made a mental note to take care of that later too.

Zach also transported a Red Angus bull back to the homeplace. They used the bull to breed a small group of cows that had late-summer calves and couldn't be bred during the usual summer period. By now the bull's job was done, but they'd had problems with him. He was roguish and prone to going through fences or jumping them. They'd gotten enough calls from neighbors about him getting out that eventually the first question they would ask was, "What color is he?" It was all they needed to know.

Del had started calling the bull Hamburger. Said he was going to sell him for hamburger. But he'd reneged on that promise because Hamburger had good bloodlines, and replacing him would be expensive. Furthermore, nobody would buy a fence jumper, so he'd go for slaughter price, nowhere near the price of a bull with his bloodlines. Del figured some guys might try to sell a bull like that at an auction and not say anything about his fence jumping, but word would get around. People knew each other in the cattle-buying business. Finally, Del had enough. He said if that bull got out one more time it was all over. He was going to sell him no matter what the price. But remarkably, after that proclamation, the bull hadn't jumped a fence in over a year.

Del was surprised. "Those cows must have kept him pretty busy if he didn't have time to think about getting out," he told Zach. "Or maybe he's getting old. But he's still on probation."

At the homeplace, Zach opened the steel corral gate to let Hamburger into the pasture with a half dozen other bulls. Immediately, he sensed a fight coming. A black bull with a neck rippling in muscles approached, head down, front hooves pawing the ground, flipping mud and dirt up over his back. Zach grabbed a bright yellow whip from behind the truck seat and ran toward them, cutting through the corral gate, his boots slipping in the muddy ooze. Before he could reach them, the black bull landed a blow. Zach stopped short for a moment, not wanting to get caught up in a fight he couldn't stop, then yelled as loud as he could and ran straight toward them, waving his arms. "HEY-AA! Get outta here! HUUUA! HAAAA!"

He brandished his whip with a loud crack. His bravado startled the dueling beasts, and they bolted like whipped dogs, but not before the black one felt a couple of sharp stings on his backside from Zach's long whip. Zach figured the interruption would be temporary, but at least they were out of the corrals where their fighting wouldn't damage a fence. Out in the open pasture they could stay away from each other. At least that was usually the case. A few years ago they bought a young bull that was smaller than the others, but they figured he had enough space to stay clear of trouble. They were wrong. The little bull got so frightened he ran out into a pond and didn't stop until he was in water so deep it almost covered his back. But even then one of the bigger bulls tried to come after him.

Over fried chicken, mashed potatoes, and hot rolls at dinner, Del, Zach, Harlan, and Caitlin discussed farm work and local news, but Del couldn't wait to tell a story about the Amish. He said when he took the exam to renew his herbicide and pesticide license, two Amish men took the exam too.

"Imagine," he said. "Here they are with their horses and buggies parked outside. Won't use a phone, or drive a car or a tractor. None of them have public electricity or running water. And they're going to get a license to buy herbicides and chemicals that are some of the most complex and advanced technology used on a farm."

"And dangerous, too, if you don't use them right," Zach added.

"It's hard to understand," Caitlin said. "But it's all about what the church elders let them do. Some groups are much stricter than others."

Listening to the conversation, Zach fidgeted in his chair, anxious to get back to work. He and Del needed to fix the leaking water valve he'd seen this morning, and he had to treat the sick cow. He also wanted to pick up some machined parts for his new pulling tractor, but when he called the shop it was closed. "Why are they closing in the middle of the week?" he complained, not expecting an answer.

On the back porch, as Harlan pulled on coveralls and boots, Del asked him to wash out the cattle trailer that Zach had used earlier in the day.

"When you finish, start cutting the rusted sheet metal off the sides and bottom with the power chisel. We'll replace anything that looks rusty or weak. It's an old trailer, but for a few hundred dollars and some paint we can recondition it almost like new," Del said, "and for a lot less money than a new one."

Caitlin had prepared food for a funeral meal later in the afternoon. The meal was a tradition at the local church, and Del volunteered to take the food to the family while Zach treated the cow with foot rot. They agreed to fix the broken float valve on the water tank afterwards.

The rectangular concrete tank with the broken float was on the rental property where Zach and Amber lived. They rented the house and had moved in during the first year they were married. The house wasn't as large as they wanted, but it was less than three miles from the Montgomery homeplace. Zach figured the convenience had to count for something.

Fixing the faulty valve was one of those little chores that sometimes turned into a full afternoon of work; a failed part might have to be backordered, or a discontinued part might necessitate replacing an entire unit. The cattle water dispenser was a freeze-proof type. It had four bright-blue plastic balls, each about the size of a large grapefruit, just a little larger than a cow's nose, and each one floated snuggly in a circular opening set into a retaining frame that covered the water tank. When a cow wanted to drink, she pushed a floating ball down with her nose to reach the water. Today the float valve controlling the water level in the tank was broken, so water overflowed the tank and formed a lagoon of icy water, mud, and manure around the water tank.

Maggie, Zach's gold and white cocker spaniel, bounded back and forth across the cattle pens with her bobbed tail wagging and her nose absorbed in a world of odors only a dog could know. Roaming through the cattle pens, she stopped to lap up some manure and urine-soaked effluent in a hoof print pockmarking the corral.

Seeing her, Zach exclaimed, "Ugh! Don't even think about trying to lick my face. Now I know why I keep you outside like other farm dogs."

Without looking up, Del added, "That's where farm dogs belong anyway."

For several weeks, Maggie had been sleeping in the big white stable barn across the driveway from Zach and Amber's house. That happened about the time they put an orphaned calf in the barn where it could be bottle-fed. There weren't any other animals in the barn, and Maggie took up with the little black calf. Other than Zach and Amber, Maggie was the only creature the calf knew, except maybe an occasional rat that passed through in the night. The dog and the calf seemed to develop a bond, and Maggie would go there several times each day. Zach wondered what the dog would do when he sold the calf. He also wondered what would happen if the female bobcat that reared two young in that barn a couple of years earlier returned. He hadn't seen the bobcat again after that one spring.

Replacing the leaking valve turned out to be simple. Afterward, Zach and Del towed a big fold-up tandem disc back to the homeplace for repair. Del planned to get Harlan started replacing the scrapers and cutters on the disc. Del had bought it cheap at a farm sale because most farmers didn't want to tackle the job of replacing the disc cutters, and the replacement cutters were expensive. But the two thousand dollars he paid for it at the auction was far below the nearly twenty-thousand-dollar cost of a new one, so even with a few thousand dollars in replacement parts, he'd have a good, reconditioned tandem disc for a fraction of the cost of a new one.

Zach spent the rest of the afternoon welding inserts into a set of rear wheel rims to widen them on his new pulling tractor. Wielding the cutting torch like a hot knife, he sliced each rim lengthwise and inserted a machine-rolled band of steel between the two halves. With everything aligned and clamped tight, he struck an arc and started welding the first insert, laying

down a series of spot welds to hold everything in place. After that, he had forty feet of welding bead to finish on each rim—more than he could do in an afternoon. The shop soon filled with an acrid volcano of blue smoke from the flickering welding arc, which bathed his helmeted profile in blue-white light. A spray of hot red slag spewed from his welds like fireworks. Some of the tiny glowing particles hit his coveralls before bouncing harmlessly to the concrete floor. He paused occasionally to inspect his progress, chipped at pieces of errant slag, then dropped his helmet again as the arc hissed and hummed and flickered blue white. Welding smoke drifted upward, finding escape through the shop's wide double doors, which were partly ajar despite the cold.

Sometime later, Harlan arrived to retrieve the power chisel and spotted Boy the shop cat squatting on the absorbent shop floor sweep compound Del has put under a tractor to soak up leaking oil. "Hey! Get away from there!" he yelled.

Harlan knew Del didn't like the cat crapping in their floor sweep, and Boy, for his part, didn't like outdoor bathroom options much. Harlan picked the cat up and unceremoniously carried him outside anyway. Boy caterwauled, but Harlan figured he'd be back inside the shop in a few minutes anyway. It wasn't the first time the cat had been in trouble.

Pausing to let the welds cool and the smoke clear, Zach paged through a trade journal looking for a bigger diesel fuel pump and larger injectors and fuel lines for the pulling engine. He scarcely noticed the commotion with Boy. The modifications, he figured, would double the horsepower with the pump turned up to maximum, but fuel pumps weren't cheap.

"Best price I've found for a high-capacity pump is sixteen hundred dollars," he said. "I been checking the Internet and parts dealers for a week now."

"If you keep spending all your money on this new pulling tractor, you're going to have to sell the one you have now," Del chided.

"I'm planning on selling it next fall or winter anyway. I just hope the engine lasts that long. It's five years old."

"I'm surprised it hasn't blown already."

"Anyway, it doesn't matter. I already got all of the pistons, sleeves, crankshaft, and rods and everything to replace it if it blows."

Zach knew he still had a lot of problems to face building this new pulling tractor and continued welding until shortly after dark when he left for a meeting of the regional tractor pullers association. He was an officer in the association and they would be reviewing rules for next year's pulling season. The members took their sport seriously, and Zach didn't want any surprises.

8

February 16
Mice and Mud

February. Majority of spring calves born this month and next; feeding cattle and assisting newborn calves are daily chores; tractor and farm machinery repair continues in farm shop; farmers attend annual Western Farm Show in Kansas City as well as local machinery and farm sales; fence construction and repair continues; great horned owls begin nesting; flocks of American robins return; chorus frogs call from wet areas on warm evenings.

Zach swung the feeding truck off the gravel road, turned up a short, bumpy lane toward a house, and then left into a large, sloping field, part of which was an old prairie pasture. The ground was soft with all the rain and snow, and he kept to a rough area where flat sandstone lay at the surface, then bounced along toward a small grove of trees atop another rocky area.

"We got mice somewhere in this truck," Zach said with resignation. "I always wonder what they're doing when I'm driving around like this, bouncing over rough ground to feed cattle."

"Hangin' on, most likely," Ross, one of the Montgomerys' landlords, responded.

"I guess so. It's hard to get 'em out of our trucks. I figure they got to have a nest in here somewhere."

Ross listened with amusement. He lived in Kansas City and didn't get out to the country much. A buzz cut muted grayness creeping into his hair, and gravity and inactivity had softened the once-youthful, farm-hewed muscles on his lanky frame. He'd known Del since high school, and Zach since he was born. When Ross's dad had given up farming, he rented the farm to Del, and later, after Ross's dad died and his mother moved to a nearby town, Zach and Amber rented the farmhouse.

Ross had tried farming for a few years after he returned from Vietnam. At first commodity prices were good. Then things changed. Government policies that had been favorable to farmers at the onset of the seventies were manipulated to hold down prices. That was when Earl Butz famously admonished farmers to "get big or get out" and scuttled or revamped many federal agricultural policies that had been in place since Roosevelt's New Deal. Ross, like many young farmers who were undercapitalized and getting started, was caught in the commodity price downdraft that followed. A lot of farmers never recovered. Ross figured he could have weathered it out, but his heart really wasn't in farming, and he left for the city. He didn't miss the endless work, or the blistering summers or cold winters. He didn't have the mechanical skills, either. He never really liked messing with greasy engines or repairing balky farm machinery. He could see that the survivors would be the ones like Del who could do those things for themselves and were good at it.

Peering through spectacles that rested a little too far down on his nose, Ross saw a large bird fly out of a tree up ahead as the truck bounced along the rocky pasture ridge. It flew with quick wing beats, not gliding like a hawk.

"Hoot owl!" Zach exclaimed, then corrected himself. "Well, I think it's called a horned owl. It's got a nest over there in that grove of oak and cedar. Flushes almost every morning when I'm feeding cattle up here. They got two fuzzy babies. One's bigger than the other."

Ross could just make out the nest. He thought it looked small, a little stick platform, maybe an old crow nest. He was pretty sure horned owls didn't build nests.

"I wonder why they're nesting in the middle of winter."

Zach didn't know, but he thought the babies were getting a head start on spring.

Fields were muddy now from the snow and rain. The cattle turned feeding areas into mud wallows, hooves churning mud. The trucks cut tracks in the muddy sod, too. That was why Zach was keeping to the higher and drier part of the pasture.

"We lost a cow in this mud less than a week ago," Zach said. "Got stuck in it. She was dead before we found her. We could see where she struggled. Everybody's saying it's one of the wettest winters they can remember."

"Farmers always say stuff like that about the weather," Ross said. "They like to complain."

"That's true, but this is unusual. The Daltons lost three cows the same way. Some guys we know lost even more."

Ross stared at the soggy field. Maybe it was true. He didn't remember anything this bad when he was farming.

Ross stayed for midday dinner with the Montgomerys. "I been gone from around here for so long I hardly know anybody here nowadays."

"Most of the ones you knew are gone," Del replied. "Caitlin and I used to know everybody living along the highway between here and Ashton. That's a fourteen-mile stretch. Today there's only two families still around, besides us, that were there when we got married. I don't know all the new ones that's moved in."

Ross pushed back from the wooden table in the Montgomery kitchen. "Country's more dangerous now, too, even out here, I guess."

"Oh, maybe a little. Nothing like what you guys got in the city. Somebody's getting killed up there almost every day."

"Everybody wants to carry a gun around, too. Used to be guys had gun racks in their pickup trucks. Now conceal and carry's the big thing."

"Nobody much putting rifles in their pickup trucks now. Somebody will just steal 'em. But I never worried about those guys, anyway. It's the ones with guns you can't see that worry me."

"People that wants them are the ones that shouldn't have them."

"Assault weapons, armor-piercing bullets, big clips. Reminds me of my Vietnam days," Del said. "People don't need all that stuff. Just doing things they shouldn't with them."

Ross, who had been in Vietnam at about the same time as Del, agreed. "Well, those were crazy days. Military probably lost more guns and equipment to the black market in 'Nam than they used in the war. Around Saigon, where I was stationed, a lot of guys were selling everything they could get their hands on, even trucks, big equipment, liquor, you name it."

"You ever had any close calls? Like getting shot at?" Del asked, changing the subject a little.

"Nah, not really. I was in a chopper once in the Mekong Delta. Engine quit on us. We auto-rotated down to a rice paddy. It hit pretty hard. Soon as that thing hit the ground, I was out of there and running for the jungle."

"You'd just be a big target sitting out there and ever'thing."

"Yeah, my buddies kept yelling, 'Hey, Ross, wait. Army will send a rescue chopper.' Well, guess what? I got to a road and made it back to the base before they did. Right after that, I managed a transfer out of the medical corps. Got a desk job."

With memories of old war stories and youthful adventures flooding back, they filed onto the porch, leaving the kitchen sink piled with dishes and the table cluttered. Outside, Ross looked with interest at Del's trucks and farm equipment.

"Takes more equipment now," Del volunteered. "Everything's bigger too. We had to double the size of all the gates so we could get into fields."

"Guess you guys forgot what a scoop shovel looks like," Ross teased.

"Oh, we got one around here somewhere," Del said, laughing. "Don't get used much. Most guys now would hardly know which end of it to take hold of."

"When we were kids we sure did. Started driving pickup trucks and tractors in fields when we were eleven or twelve, some even younger."

"That's changed some," Del said. "Equipment's so big and expensive now you don't see kids turned loose operating that stuff like they used to. Farm equipment's more complicated now too. Besides, not so many farm kids around anymore. Zach was the only one in his high school class to stay on the farm."

Ross and Del walked out to the farm shop where Zach was already at work. "We got to start hauling lime, spreading it on the fields if they ever dry out," Del said. "Need to pregnancy check some cows, too, but I hate to work in this mud. Corrals and ever'thing gets so muddy."

In the shop Boy arched and rubbed against Ross's leg. "That your cat?" Ross said, looking down at Boy, who was already moving off.

"Yeah. He's getting to be an awful high-priced cat. About died on us last fall. That was the second time we had to take him to the vet. First time he got a urinary tract infection, then a few months later he got an infected paw. That's when we had him neutered."

"Doing all that was pretty expensive, wasn't it? Is he . . . like a pure breed or something?"

"Nah, just a farm cat. But I hated to see him suffer. I told the vet we might as well just fix both ends at the same time. I didn't want to have to bring him back. I was getting tired of him squirting on my shop tools, anyway, actin' like they were his property and ever'thing."

"He looks fine now."

"Oh, sure. Vet gave us some antibiotics. I just held him up and Caitlin injected it. Made a lump under his skin, but he absorbed it after a while. He's fine now."

Ross left in the early afternoon, just as Harlan was arriving, their vehicles passing in the driveway. Harlan parked his battered old pickup alongside one of the shiny steel bins and climbed out, muttering something about not waking up, then headed straight to work in the shop where Del's International Harvester 915 combine was in the midst of annual winter service. The 915 was an older combine, although once it had been the largest one built by International Harvester. Of all the machinery that went through maintenance each winter, the three combines were the biggest jobs.

This year the 915 came with a stowaway. Zach got the combine out of a winter storage building about a mile away. When Zach drove it inside the shop, Del saw a stray cat leap out of the rear of the machine. The cat fled to the back of the shop and crouched under a pile of steel and angle iron and tools. For two days they tried to get the cat out of the shop. It didn't seem wild, just frightened. Sometimes it would even sit up on top of the combine, watching, but they couldn't touch it or catch it. Finally, they enticed it outside with food. Then it disappeared, even when they left food for it.

"That cat was luckier than the other one," Harlan said. "Remember a few years back when we had another combine parked in that same barn?"

Del said he remembered all too well.

Harlan had backed the combine out of the shed and, with the engine idling, put the threshing mechanism in gear and walked around the combine to inspect the moving parts and clean out the dust and chaff. Satisfied, he'd revved the throttle and brought the machine up to normal speed to shake out any remaining debris. That was when he heard a dull thud at the rear of the combine. He shut the machine down and went around back to check on the sound, unaware there had been a stray cat in the straw walker bay at the back of the combine. The cat overstayed, or maybe it had been too frightened or confused to escape when the machine was in gear and idling. At full throttle Harlan guessed it couldn't stay ahead of the straw walkers, which would have quickly bounced everything rearward. He figured the cat just got "walked" or pitched out the rear of the machine. He didn't know. Maybe the cat panicked and leaped out. Unfortunately for the cat, that combine was equipped with a straw chopper designed to mulch the straw and spread it out behind the machine.

"One leap and that cat used up all his nine lives right there," Harlan said, shaking his head. "Man, what I saw back there wasn't chopped straw. I felt real bad after that. Almost made me sick. Still bothers me thinking about it."

"It was a shame, really," Del added. "That cat was probably doing us a favor catching mice inside the combine and hanging around that old building."

"Wish a cat would hang around that truck I was using this morning," Zach said. "Maybe do me a favor and get rid of those mice. No way to clean all the seeds out of those trucks."

"Combine sittin' around is like a magnet for mice, too . . . and for any stray cat that thinks it's found paradise with all the mice in there," Del added.

"Well, that one did, I guess," Harlan said, a frown crossing his tanned face. "Couple years ago, when I got one of these combines out of winter storage and started it up, it threw an old, dried-up rabbit pelt out the back. Guess a cat musta dragged that up inside the combine too. Ain't no rabbit gonna jump up in there."

The men got to work with Del, climbing up to the top of the combine and getting down inside the huge grain tank to repair a worn area where the grain exited into the auger. Down inside the big tank he called out instructions to Zach, who stood on a stool outside the bin. From inside the bin, Del's voice echoed upward. "Okay, I'm pushing the bolt through, can you see it?"

"Yep. Give me a second to get it fastened," Zach responded.

They worked, each blind to the other, one outside, the other inside, hollering instructions to each other in voices altered by the metal architecture of the big grain bin. The auger pipe was bent where it met the grain tank and the Archimedes' screw inside had worn a hole in the bent pipe. They straightened the pipe, bolted a piece of metal over the hole, and welded it solid. They'd been welding and replacing parts on the combine for several days now. But the biggest, most vexing job lay ahead. The main serpentine drive belt powered half the combine's moving parts, and it was pinched and torn and wouldn't last another season. Some fourteen feet long, the belt looped and doubled back around a number of pulleys on the right side of the combine. Some of the pulleys were positioned behind other obstacles and were difficult to access. Zach and Del worked for two hours to free up pulleys, chains, variators, and shields so they could replace the belt. Zach wiggled up under a huge metal shield on the outside of the combine and next to the clean returns elevator, and then fastened a small portable light to a support so he could see. Struggling, muttering, their conversation diminished to grunts and monosyllables.

"Wait a second. Don't push the belt off 'til I get hold of it."

"Got it now?"

"Nope."

"Reach behind the returns elevator."

"Can't."

"Here, let me drop this chain off the sprocket first."

"Okay. I got it. I see the bolt that's in the way, but there ain't no threads on it."

"What do ya need to do to get it off?"

"Don't know yet."

Pausing to rest, Del exclaimed, "I know a couple farmers that had to get factory mechanics to come out and replace this same belt on a John Deere combine. Takes two people to change it. Cost them over three hundred dollars just to change this one belt . . . I can almost see why."

They worked through the afternoon, once starting the engine to rotate a pulley wheel into better position. The shop filled with diesel smoke, which swirled up and gathered in a caliginous cloud in the rafters. Eventually they pushed open the big west-facing bay doors a little to clear the choking air. A cold breeze rushed in, and a bar of afternoon sunlight flooded the front of the shop and made it seem warmer than it really was.

"Not as smoky as it was the other day when we started that 1568 diesel tractor in here," Zach said as Del stood in the doorway.

"That was bad. Filled up the shop with so much smoke I could barely see to get out the door."

By late afternoon, with sunlight reaching the rear of the shop and their bodies casting long shadows across the shop floor, they pronounced the combine ready for a new season, backed it out into the driveway, and engaged the threshing mechanism, the engine idling. At slow speed the machine rumbled in a kind of controlled chaos of rattling and shaking. For several minutes they walked around the great red leviathan, pausing to watch and listen for abnormal sounds, occasionally exchanging a few words as belts turned, arms and pulleys rotated, and chains and sieves scraped. Wisps of grain dust and chaff seeped from tiny openings at the edges of the machine's thin skin of sheet metal. To an onlooker it would have seemed a chaos of noise, but Del and Zach heard every moving part, not disconnected, but as finely tuned parts of a whole, almost like a conductor hears the instruments in his orchestra. Everything—reel, auger, threshing cylinder, chains, return elevators, sieves, walkers, even well-oiled bearings—had its distinctive sound and were part of the overall harmony of the machine. Eventually satisfied, Zach climbed up the skeleton stairs to the combine's cab and shut the engine off. The noise stopped, not abruptly but slowly, as hundreds of moving parts scraped and rattled and slowly rumbled to a protesting halt. Out of the silence the dust cleared and the great machine, tuned and adjusted, was ready for an orchestral harvest of its own.

The next day, after an early dinner, Del, Zach, and Harlan set up a temporary corral from steel panels that hooked together. Each year a few calves were born during the summer and early fall months, too late to include with the November sale of spring calves. These late arrivals were from cows that didn't breed quickly or that were on different breeding schedules when Del purchased them. They usually had to be weaned and sold in the spring, and it meant an extra off-season roundup—a lot of work to separate a few cows from their calves.

"Might as well do this all at once," he said to Harlan and Zach, who were already hitching up cattle trailers and loading an all-terrain vehicle into the back of a trailer.

Harlan said he felt like a gasoline-powered cowboy. "Everybody's bucking across pastures on these rubber-tired ATVs now. Chasin' cows with these things instead of horses ain't so bad, though. Saves me some running. I hurt too much to run. Can't do that no more. And I sure as hell don't want to ride no horse."

Del told Harlan to stand in the back of the funnel-shaped corral, close to the exit. "Watch this gate, Harlan. Don't let any cows through unless I give the signal."

Del, in rubber boots, was already ankle deep in mud in the temporary corral. He studied his herd book, which showed ear tag numbers, calf tag numbers, and the birth dates of each. There was a long pause. Finally, looking up, he said, "All right. Let's get the cows with white ear tags outta here. But no yellow tags. We'll have to sort them later by tag number. I'll tell you when we get to them."

"Okay! Just let me know," Harlan hollered. "You gotta tell me which ones to cull out 'cause I sure don't have no idea which ones to let through. I'm just like a damn mushroom out here, you know. Kept in the dark and fed a bunch of crap," he said, grinning. "Only right now I'm just standing in a bunch of crap."

If Del heard Harlan, it wasn't apparent. He was already moving through the milling herd, arms outspread, talking softly to the cattle, poking his bright-yellow whip in their faces to halt escape attempts. "Yup, here's one!" Del yelled. "Let her go!"

"Got her," Harlan responded, opening the gate just enough to let the cow run through.

"Watch that one!" Del yelled. "Don't let her through."

Within ten minutes, Zach and Del, moving among the mud-spattered herd, reduced the cow numbers by half. Harlan whistled and hollered to halt escaping cows, then yelled again to encourage the culls to go out the gate as he dodged flying mud and hooves. As more cows were culled, it became harder to separate the remaining ones because now most of the cattle in the pen were holdbacks. Also, being continually pushed around in the corral, they were getting wise to the men's intentions. Del continued moving slowly back and forth. With his mud-stained yellow whip, it usually took only a sharp rap on the nose or a stinging whack on the butt to dissuade even the stubborn ones, but the slippery footing was treacherous for men and animals alike.

"Like a swamp in here, ain't it?" Del yelled.

"More like a mess of cow crap," Harlan muttered. "Just staying upright is a full-time job."

Zach bunched a dozen cows and their calves at the far end of the corral, which narrowed to a bottleneck. Closing a gate behind them, the only avenue of escape for the trapped animals was up into the trailer. He yelled and rapped his whip against the metal corral panels, creating a sense of urgency. A lead cow hesitated, then pushed by others from behind, made the little jump upward into the trailer. Encouraged by her "escape," others quickly followed, rushing headlong into the trailer. Zach grabbed the trailer's tailgate rope and jerked it free, and the gate slammed shut with a loud rattle.

Del and Harlan continued to sort the last holdouts. Their boots and coveralls were now mud-splattered top to bottom, and the cattle were also muddy. It had taken an hour to load the two trailers. But that was only the beginning. The truck wheels spun out, leaving shiny ruts in the mud. They barely budged the loaded trailers. Eventually, the trucks, with trailers in tow, had to be dragged onto solid ground with a tractor. More time lost. Del groused that it took almost as much time to get the trailers and trucks out of the mud as it did to load the cattle.

They had to repeat the corralling and sorting on another farm two miles distant where there were another half dozen late summer calves. They'd need only one trailer for that group.

At the homeplace, beneath the fading red glow of a winter evening, the men continued their work, this time separating the calves from their mothers. A swirl of grackles, their plumage burnished and shot with purples and golds in the evening light, funneled across the pasture south of the buildings. In a long, undulating column they looped up over the barn and corrals, then swooped down and fanned out into a stubble field just north of the buildings. Like a brash, rowdy mob, they set to work immediately, strutting and chattering noisily as they foraged across the field. Del guessed they might be after spilled grain from the harvest last fall. The air was cold, not like spring yet, but the grackles' sudden appearance hinted of a new season. There was already a tinge of green in pastures. Or so it seemed. He suspected his cattle had noticed. By the time Del and Zach trucked their calves to the auction barn and returned home it was after midnight.

9

February 26
Nursery

It was still dark when Del climbed into his feeding truck. The engine grumbled. A tower of white exhaust rose in the blue-cold predawn. He eased the truck into the frozen corral and squinted at the white steel-pipe fence in the truck headlights and at the herd of bred heifers penned there. The fresh snow in the corral was already a churn of hoof marks. "Hard to believe in less than six weeks we could be planting corn," he mumbled to himself.

He stopped the truck beside a metal feeding trough that resembled a thirty-foot-long steel drum sliced lengthwise and laid on its side. Its metal supporting legs were held fast in frozen manure. At this moment, it would have been difficult to find another living thing on the farm that shared his optimism for spring's return.

Stepping out of the truck cab, Del scraped the snow out of the feed trough with a flat-bottom grain scoop. Opening the truck door, he climbed back inside and flipped a hand-held electric switch that controlled a short auger on the side of the truck. A stream of dusty white ground milo poured into the trough. Smiling with satisfaction, he let it run a little longer than usual. Immediately, a long row of black faces and bodies queued up, jostling for space alongside the trough. Their moist, steamy breaths sent little plumes of fog and grain dust into the air, and the cows made noisy puffing sounds as they pushed and shoved and licked ground milo off wet noses with rough tongues. These were Brangus heifers, a reproductive aristocracy-in-waiting, each nearly identical, black with thick, heavy coats, deep chests, and moist dark eyes set in broad faces. In the truck lights they looked like a regiment of clones in black, mirror images resolving into alignment alongside the feed trough.

At that moment, something dark at the far side of the corral caught Del's attention. A heifer was standing alone. Then he saw the calf, the first from this generation of heifers. The little black calf was still wet but already standing on long, wobbly legs. It pressed against its young mother.

"Good girl. Lick it and get that circulation going," he said out loud.

Moments later, Caitlin crossed the driveway. Bundled in a heavy coat and scarf, she carried a bottle of formulated milk for an orphaned twin calf.

"First heifer just had a calf," Del yelled to her. "This is one we won't have to pull."

Caitlin could sense the excitement in his voice. Heifers bred to large bulls often had trouble calving, and it was always a risk having to manually pull a calf. A couple years ago, he'd read about heifers being bred to longhorns so they would have smaller calves, and it was an idea he decided to try. He bought a longhorn bull and bred his heifers to it. The hardest part had been finding a black longhorn bull. Most of them were spotted.

Caitlin and Del had been up for more than an hour checking weather reports over breakfast and talking about the birth of their first granddaughter a few days ago. They had already visited their daughter Wendy and her new baby, and Caitlin was so excited she wanted to see her granddaughter again. Del might not have admitted it, but he wanted to see the baby again too. Still, when the weather was bad he found it difficult to think about anything that would take him away from his cattle for more than a few hours. On days like this, he would start chores even earlier than usual.

If the weather had been good, he might have spent a little time working on farm records before breakfast. He liked to organize farm bills and receipts before Caitlin posted them. They posted receipts and expenses in a wide farm ledger book. There were columns for hired labor, repairs and maintenance, interest, rent, feed purchases, seeds and plants, fertilizers and chemicals, machine hire, supplies, breeding fees, veterinary and medicine, fuel, warehousing and storage, taxes, insurance, utilities, freight and trucking, conservation, principal paid on notes, and several columns for new farm investments, and more for farm and non-farm tax deductions. The record-keeping seemed endless. Caitlin also would enter the records into a computer, but Del didn't entirely trust computers.

"What if something happens? Maybe the electricity goes off. Then poof! Everything's gone," he said. "Then what?"

"Then you start over," Caitlin said, teasing him. Del didn't object to the computer, but he insisted on a paper backup of everything.

During calving season from late January through March, Del always left the house before first light. Some farmers started winter mornings with coffee and gossip at the diner in Prairie Point. They arrived early, too, usually before first light. Most of them didn't stay long, but old timers and retirees lingered. In the past Del occasionally went, but it took up too much time. He'd told Caitlin, "Most of what those guys talk about I don't need to know. Besides, I don't even like coffee."

Caitlin hurried over to see the new baby calf. Del was bubbling with excitement. Anything that reduced calving problems meant fewer orphan calves to feed, and hand-reared calves were a lot of work. The orphan calf she was feeding took the bottled milk quickly. On her way back to the house, she nearly stepped on Boy the shop cat squatting in a tire track in the snow. He leaped aside and landed unceremoniously in a snow bank, almost disappearing from view. Caitlin glanced at the cat but she was already thinking about her work at the post office and planning how she could fit in another visit to see her new granddaughter.

Del had observed that cows seemed more likely to have calves when the weather was bad. He didn't know why. Taking care of cattle was a little like midwifing except for being out in the weather. The day before temperatures had been near zero, and he had counted ten new calves. By the end of the day, he'd brought six of them in to warm up and dry, putting them temporarily in the barn and shop. Sometimes he marveled that any survived, being dumped wet and shivering onto a snowy field in freezing temperatures. Inevitably, there would be birthing complications, cows that wouldn't claim their calves, cows that didn't have enough milk. At times he could help, but the weather was often the final arbitrator of life and death.

As the dawn sky reddened, a collusion of crows, like lumps of coal in blue snow, gathered in the pasture beyond the corral and feedlots. They had spied the heifer's afterbirth, a messy pool of blood-streaked gray tissue at the side of the corral. A few horned larks, small birds with black bars framing yellowish faces, perched on fence posts or loitered in bare patches near the snowy corrals. Several meadowlarks gathered near the feedlots. The presence of so many birds was a sign of stress. On the coldest days they sought food around the farm buildings where they looked for spilled grain, seeds in manure, and dried insects in the hay. When temperatures warmed and the snow disappeared, they would disperse. Horned larks were among the hardiest, sometimes nesting on bare hard ground by the end of February.

Del swung his truck onto the highway, drove a half-mile west past snow-covered fields, then turned south onto a gravel road. It was 6:20 a.m. By now there was a soft red glow in the eastern sky. Dark fence posts stood in sharp relief against blue-white fields. Not a vehicle track marred the snowy roadbed or the fields beyond. Three-quarters of a mile ahead, he turned left onto a short lane leading to one of the farms he rented. He stored part of his winter hay supply here—giant round bales arranged in long, orderly rows. At this time of year, the shaggy bales were dry and bleached on the outside, but the hay was still fresh inside.

The house that had once stood on the farm had been torn down years ago. Two metal-covered barns and a projecting wellhead covered by a bucket were all that remained. The landlady, in her nineties, lived in the east with a daughter. As far as Del knew, neither had seen the land they'd inherited. Even so, the old lady took a sharp interest in her property. She scrutinized expenses and didn't hesitate to call the feed store or Del if she didn't understand a billing. Del and Caitlin had had some laughs over her vigilance. Once, she questioned his fertilizer bills. "Sonny," she'd told him, "I don't think you're using enough fertilizer. I don't mind spending money for fertilizer, and I'm not going to waste my time and money with you if you don't try to grow good crops."

Del had to explain to her that he had started applying nitrogen in the form of anhydrous ammonia, something that earlier generations of farmers weren't able to do, and he was applying just as much nitrogen as always. She had noticed the change because her portion of the fertilizer cost was lower. He would have expected the opposite from most landowners.

Hoisting two huge bales aboard the truck with the hydraulic lift arms at the back, he gunned the vehicle back out onto the road. Two nights ago a calf, only a few days old, got out on some slushy ice on a pond and couldn't get up, its legs sticking out to the sides. It froze hard that night, and when Del found it the next morning its legs were frozen fast to the ice. He chopped its legs free, brought it home, and stood it up with its legs, still encased in ice, in four buckets of warm water. Last night the calf's feet were swollen but it was walking. He reckoned it would lose its feet.

As he drove to the largest cattle herd, Del met Zach. The two stopped briefly in the middle of the road, their snowy truck tracks separating like rail lines in a switching yard. With windows down and little clouds of condensation forming each time they spoke, they discussed morning work plans. Del told Zach about the calf with the frozen feet.

Zach shook his head. "All we can do is eat him, I guess."

"We'll have to butcher him. Put the meat in the freezer. Nobody will buy him without feet."

"Won't taste any different than any other," Zach said. "We can use some tender veal steaks."

Del discovered three more newborn calves in the next herd. He had often told Caitlin that finding newborn calves was like finding money. But Caitlin, always cautious, had said, "Yes, and just as hard to hold onto." They both knew that keeping calves healthy until they turned into money was the challenge.

Del stopped the truck in the cow pasture, lowered one of the bales to the ground, and then, by driving forward, unrolled it like a fluffy, hundred-yard-long mattress atop the snow. He knew the cattle would waste a lot of it, but the new calves would have a warm place to lay off the snow, and he was more concerned about saving calves than a little hay. About a third of the cattle in this herd were Charolais, a breed developed in France several hundred years ago but seldom seen in the United States until the mid-1960s. Charolais had grown in popularity because they were large and gained weight rapidly, and there was no denying their visual appeal. Their calves' creamy-white coats, black eyes, and pink noses were just like their mothers'.

After feeding each herd, Del always checked streambanks and areas away from the herd for newborn calves hidden by their mothers. He found one on a snowy slope a quarter mile from the main herd. It wasn't unusual for cows to distance themselves from a herd when they calved, but sometimes the calves slipped down streambanks or fell into icy water when their mothers moved them. This one looked cold. It wasn't standing—not a good sign. Del decided to leave it for now. He wanted the cow to get the scent of her calf firmly imprinted. Even if it was weak, it would survive a few hours.

At 8:50 a.m., while feeding the last herd, Del's cellphone rang. It was Zach.

"Dad, got a cow down in a pond over on Dunstan's. Can you bring over a strong rope? Mine just broke when I tried to pull her out."

It was the first bad news of the morning.

"I'm on the eighty east of Prairie Point. Take me about fifteen minutes to get there," he said in a matter-of-fact tone. He knew that if a cow broke through ice, or got mired in mud, there was a fifty-fifty chance she wouldn't make it, but cattle could survive exposure like this much longer than humans.

Del stood at the edge of the pond and looked at the cow, a smallish black animal with a partial white face. He thought she might be an older cow. He shrugged. This happened occasionally, and there wasn't much they could do. The luckless cow had walked out onto the ice, and now her body was half submerged in the icy pond, her head and body pointed toward the center of the pond. She would be difficult to pull. Del backed his truck off to one side and close to the pond's edge. Zach stepped gingerly onto the ice, as close to the cow as he dared without breaking through the ice himself. He could see her large dark eyes bulging with fear and confusion. Forming the rope into a lasso, he swung it over her head and threw the opposite end to Del waiting on shore.

Del eased the truck forward, four wheels biting snow. The rope tightened, swinging the cow's head and neck around to the left. Her neck stretched tight, extended almost grotesquely. Zach could hear vertebra making popping sounds. Her inert body didn't budge. Del stopped the truck, unsure if he should proceed. The cow's neck and head were stretched to sickening proportions. He worried that the tension might kill her outright.

"Keep going," Zach yelled, motioning frantically with his arms. "Don't stop now!"

Del continued, edging the truck forward, inch by agonizing inch. Suddenly there was a gasping, sucking noise.

"Here she comes!" Zach yelled. "Keep pulling, just go easy."

Slowly the cow's tortured foreparts moved, and there was a loud sucking sound as the animal slowly rose from her ice- and mud-filled tomb and rolled onto her side. She slid onto her back on the frozen shore, legs skyward, neck extended like a giraffe. Del continued dragging her a short distance up the snowy pasture.

"Pull her a little farther up the slope so she doesn't slide back into the pond," Zach advised.

The cow slid easily in the snow on her back, and Del brought the truck to a halt on a gentle slope twenty yards above the pond. Zach unfastened the rope, and the cow immediately rolled over into a resting position and raised her head. Her front legs were folded beneath her body. She seemed unhurt and was breathing well but didn't attempt to stand. Del and Zach stood for a moment, relief in their faces. They looked at the cow. Her trauma wasn't obvious in her demeanor. Del wished there was some way to communicate with her.

After a long pause he asked, "She have a calf?"

"No. Not yet, but she's due any day."

"Too bad. She may not make it. We could lose her and her unborn calf."

"All we can do, really, is make her comfortable," Zach said.

Del was pragmatic about the outcome. "You can't tell by looking if she's hurting inside or not. I'll unroll some hay off the truck. You push it around her."

Together they managed to roll the unfortunate cow onto a layer of fresh hay, then tucked more around her, even piling some on her back for a little insulation.

She reached for a mouthful and began to chew, a good omen.

"No telling if she will live or not," Del said. "She may live a few days. As often as not they eventually die."

"Well, for sure she would have died in the pond."

Back at the homeplace, Del looked in on Harlan's progress in the machine shop. He was replacing throw-out bearings and a clutch pressure plate on their big International 1468 tractor. Harlan had the front and back halves of the tractor suspended by wheeled supports and overhead chains. A portable light illuminated a glitter of shiny gear parts inside a cavernous, oil-slicked transmission casing. Harlan said he needed help reconnecting the two halves of the tractor. "Don't have enough hands to do it myself."

"We'll be back in a few minutes," Del said, already on his way out the shop. "Zach and I got to pick up a new calf that's weak. We'll get this tractor bolted together again soon as we get back."

For the first day or two of life, newborn calves are too uncoordinated to run well, so catching them isn't difficult. Staying out of harm's way of an angry mother cow, however, is another problem. Del drove his truck alongside the weak calf he'd seen earlier in the morning. He jumped out, scooped up the calf in his arms, and hurried around to the passenger side of the truck.

"Quick! Scoot over in the middle. Let me push the calf in here on the floorboards."

"Got him," Zach responded, holding the calf down with a foot although it was so cold and stiff it didn't struggle much.

Del moved so quickly that the cow didn't catch on immediately. By the time she came after him, he was already back around on the driver's side of the truck. "No room for hesitation," he said, pausing to get his breath. "Some of these ol' gals will really try to fight you. This one wasn't too aggressive."

Zach grinned. "Didn't know you could move so fast."

When Del and Zach found a weak newborn calf, they usually brought it in for a few hours, or sometimes overnight, to warm up in the barn or shop. If a calf appeared unusually weak, or cold and stiff, they put it in a large wooden box with a heat lamp and fan at the top. Del gave rehydrated colostrum to the weakest ones, but it was expensive. He always tried to get them to nurse on their own first. Cows produced colostrum instead of milk only for a couple of days after birth, and calves absorbed it best in the first few hours. After that it didn't do them much good because they lost the ability to absorb the antibodies in it. But without it calves often didn't thrive.

Del had saved more calves than he could count by bringing weak ones in for a few hours to a warmer place. The biggest risk to a newborn calf was freezing rain. When that happened, they could die in an hour or two if he didn't find them. He'd learned, however, that cows needed time to memorize the scent of their calves, so he tried to leave the calves, even weak ones, with the mothers for a few hours. Once the scent was imprinted, a cow would remember the scent even if he took the calf away overnight. He was particularly careful, however, not to mix the scent of one calf with that of another. When he brought in two calves, one always rode on the truck floorboards, the other in the seat; and if he put a calf on a burlap sack on the shop floor, he wouldn't reuse it for a week or more so the scent could wear off.

Scent was so important in offspring recognition that when a cow lost her calf, the best way Del had found to get a cow to adopt an orphan calf was to tie the skin of her dead calf onto the back of the orphan. After a few days, with scents intermixed, cows would usually claim the orphan as their own.

Back in the shop, Del and Zach helped Harlan ease the two halves of the tractor together. With wheeled supports underneath the front half of the tractor and safety chains attached to it from a sliding track above, they maneuvered it backward until it fit snuggly against the rear half, engine mated to rear end once again. Harlan then began the laborious task of reconnecting everything.

"We just saved ourselves at least a thousand, maybe fifteen hundred dollars," Zach said as he surveyed Harlan's work.

"Yeah, but you gotta pay me," Harlan grinned, wiping the back of a greasy hand on his pant legs and looking up from beneath the tractor. "So it ain't quite free."

Caitlin hadn't returned home from morning work at the post office yet, so Del and Zach figured they had time to take two of their trucks to the inspection station at Spencer's Truck and Tire near Prairie Point for annual license renewal. Inside Spencer's heated garage, chunks of dirty snow and ice frozen to the undercarriages and fenders began to thaw and fall on the garage floor, adding to a growing lagoon of oil-slicked water spreading across the inspection bay. Stepping around the worse of the pools of snowmelt, Del walked into the office to chat with the garage owner while Zach stayed with the mechanic who did the inspections.

On the way home, Del noticed a truck driving slowly along a snowy side road. Lowering the window, he could hear dogs baying in the distance. Coyote hunters. There had been a resurgence of coyote hunting, and he wondered if the state was paying a bounty again. He thought coyote hunting had always been more about camaraderie and baying hounds and maybe a little whisky drinking than money. He didn't mind the hunting, but if coyotes ran through herds of cattle to confuse the dogs, the noise of a bunch of baying hounds could frighten cattle. However, most hunters nowadays had transmitters on their dogs so they kept better control of them than in the past.

Some people thought coyotes had increased after the conservation department stopped paying a bounty, and they wondered if the disappearance of prairie chickens in Missouri was linked to increased coyote numbers. However, prairie chickens had been in decline since the early part of the nineteenth century when unregulated hunting and the loss of tallgrass prairie to the plough likely sealed their fate. There were ups and downs, but by the last half of the twentieth century, prairie chicken populations in Missouri were fragmented and in dramatic decline. Where once there were hundreds of thousands, maybe millions, there were now only a few hundred. Even reintroductions largely failed.

This was the time when prairie chickens would have begun spring courtship. Their hallmark booming, a low, hollow "uu-aaaaa-uoo" once could be heard a mile away on crisp windless mornings. Del hadn't heard them since the early 1990s. He remembered the way the chickens used to flush, dozens scattering across the rolling winter landscape in long glides interrupted with occasional furious flurries of wing beats. Their courtship dance was so impressive that American Indian dancers copied their movements, but the dance of the prairie chicken was now mostly a memory in Missouri—to younger generations not even a memory. It is as if the birds never existed. Some called it generational amnesia. Zach hardly remembered them. Del has seen this icon of the tallgrass prairies disappear—now a ghost of prairies past, a heritage that his grandchildren and future generations would probably never see again in Missouri. It was hard to put a price on its loss, but he wasn't sure what could have been done. Conservationists were still attempting re-introductions, but the birds were caught up in changes bigger than themselves, bigger than the farms and families that settled here. The prairie chicken, and so many other species, were victims of an agricultural revolution so immense and so complete that it has changed the land forever in less than two centuries.

Del pulled the newborn calf out of the warm box in the barn. Its coat was dry and soft, its eyes bright. As he cradled it in his arms, carrying it to the truck, it flexed in his arms, a good sign. Placing it on the floor of the truck cab, he drove back to the calf's mother. She was still standing where she had given birth. Del didn't give the calf to her. Instead he wanted the cow to follow him to the cattle herd. The calf would be safer there.

He leaned out the truck window and gave a few low calls. "*Ca-a-a-a-f . . . ca-a-a-a-f.*" It was the same call he used when he fed supplement to the cattle. Perhaps also the cow caught the scent of her calf, but in any case she followed him to the herd. Del put the calf on some hay and backed the truck away. The cow immediately went to her calf, looked around protectively, and began licking it. If the calf stayed on the hay, insulated from the snow or cold ground for a few hours, it would probably be fine.

After dinner, Del, Zach, and Harlan gathered in the farm shop for the afternoon. Zach worked on his pulling tractor. He desperately wanted to have it ready for some test pulls before the pull season ended.

At four in the afternoon, Del and Zach left the shop to make cattle rounds again, this time checking mainly for new calves. They seldom finished before dark. Two days before, it was nearly eight p.m. before they finished. Their nursery was growing.

10

March 6
Orphans

March. Newborn calves arrive daily; farmers feed cattle, assist newborn calves during cold weather; maintenance work continues in farm shop; fence building and repair is ongoing; by end of the month cattle seek fresh grass; wild onion and garlic appear in pastures and woods; gooseberries and wild plums bloom along woods and fencerows.

Somewhere in a cold, bare pasture in March, a baby calf is born. Instinctively its mother licks it to dry its coat and stimulate its circulation. She nudges it gently to encourage it to stand, wobbly-legged, and to nurse. It is unsteady and uncomprehending in its new world, but that changes quickly. Its first summer will be spent in carefree abandon, growing rapidly, secure in the sphere of its mother's watchful eyes. It will be one of more than fifty million born each spring in pastures across North America. But, for a few, something goes wrong. A difficult birth, an accident, a mother lost, a calf abandoned for unknown reasons. These are the orphans, and their fates are uncertain.

The manure-stained mud in the cattle pens beside the barn was frozen. As Del backed the truck to unload a huge bale of hay, tires scrunched and flexed against the hoof-marked surface of the feedlot. The steel arms of the truck's bale loader separated, and the bale fell heavily into the circular steel bale holder. Del's eyes roamed over the stocky Black Angus and black Brangus heifers. Six bulls, having made temporary peace among themselves, loafed in the largest corral with a gate open to a pasture. They had chewed their way through half a bale from a previous feeding. Another corral held a prolapsed cow and her newborn calf. Two days before, the cow's swollen vagina extruded in a sickening manner. Del called the vet. He delivered the calf, then pushed the cow's pink reproductive innards back inside and sutured them in place with shoestring-sized cord. Del would remove the sutures in a few days. He marveled at the resistance of cattle to infections.

Coming across the driveway, Caitlin carried three quart-sized bottles of freshly prepared calf formula. She was feeding three orphan calves, up from one a week ago. The three queued up at her approach, pushing against the wooden enclosure, moving their jaws and making suckling sounds. Two were twins that had been separated from their siblings because beef cows seldom had sufficient milk for two calves. Whenever a cow had twins, Del usually removed one, and he and Caitlin would then hand-rear it. It improved the odds of both calves surviving. Hand-reared twins would know only Caitlin's voice and a rubber-teated, plastic milk bottle. The third orphan was from the cow that had broken through the ice in the pond the month before. She never stood or walked again, but had attempted to give birth a day later. Zach saw the calf's fluid-stained, yellowish feet protruding and the slow birth progress. Sitting on the ground behind the cow, with his boots braced against her hips, he'd grasped the slippery front legs and pulled, at first gently, then with more force. He was surprised when suddenly the calf, wrapped in its shiny birth sac, slid out onto the ground right between his legs. But the cow was weak. Spent from the trauma of the cold water, the rescue, and the birth, her strength failed. A day later she died.

Del towed her large, stiff body to a wooded ravine. He didn't like leaving carcasses in ravines, but he had few options now. When he was a kid, animal rendering services picked up dead animals free of charge. Even as profits declined, they continued, asking only a modest fee of ten or fifteen dollars. If farmers weren't around, they left money for the renderers in a glass jar near the animal. Eventually financial costs made the service untenable, and rendering services in rural areas disappeared. It wasn't a major concern to some cattlemen because there had always been questions about disease transmission associated with the trucks and transport of dead animals on and off farms. Del and his neighbors had not had an easy method for disposal of large animal carcasses for more than two decades. Sometimes they could add the carcasses to piles of brush and burn them, but the carcasses were often wet and difficult to burn. Some farmers just let nature take its course.

Rendering had always operated on the fringes of polite society. No longer serving rural areas, renderers are now centralized near large cities and meat processors. Half of every cow and a third of every pig butchered in this country are ultimately sent to a renderer, as are millions of pets. Rendered fat can find its way into an astonishing array of products from cosmetics to paints, even reprocessed as an additive in pet foods. Yet the central issue is disease transmission and how to control or restrict certain products so that they are not fed to animals destined for human consumption. Mad cow disease dramatically changed economic equations in the rending business. It brought much-needed transparency to an industry saddled with controversial practices. As a result, feeding cattle a bone meal supplement derived from other ruminants is now prohibited. But rendered cattle products are still incorporated into products fed to hogs, which means that near-indestructible prions, a type of abnormal protein implicated in mad cow disease, can remain in the biological loop. Animal rendering remains something of an American "untouchable," a "black hole" mostly out of sight, hence out of mind, and industry safeguards have evolved slowly.

During the more than two-month-long calving season, cattle chores followed a routine altered only slightly by weather—cold temperatures, snowstorms, freezing rains—and by muddy conditions. Del and Zach were especially vigilant during cold winter rains. Although they generally split their duties, both of them kept logbooks in their trucks with the histories and health of the cows and calves in all the herds, every animal identified by a numbered ear tag. Yellow tags were Zach's cows, white tags Del's. Tag numbers were associated with birthing dates, year by year, for each animal, as well as the amount of hay fed to each herd, and dates when herds received supplements. Young calves were particularly susceptible to scours, an acute diarrhea infection. To keep track of calves treated for scours, they ear-tagged or marked them with a large number on their backs using a cattle-marking pen. Del said it was kind of like walking around with your medical history on your back, or at least on your ear tag. Caitlin thought they looked funny, but Del said, "Well, it don't matter much what's on their backs. Cows can't read anyway."

Harlan was gone, and the shop was quiet without him. He had asked for the morning off so he could sell one of his own calves at a sale barn. He borrowed a small trailer from a neighbor but it was unlicensed. That didn't concern him much, but when a tire went flat his request for the morning stretched to a full day. A week earlier he'd lost a day of work when the plumbing froze in his house, which sprayed water over the bathroom when it thawed the next day.

"And the week before that it was something else," Del noted with resignation. Harlan lurched from crisis to crisis but was unfailingly cheerful and upbeat, inevitably bouncing back from adversity. He shrugged off life's roadblocks with resigned equanimity. He'd never really gotten his life organized and was often late to work. His marriage was rocky, and he liked to fish a little too much, but once at work he didn't slack. He'd stay late to make up hours.

After morning chores, Del and Zach reconvened in the machine shop as they did most winter days. Del was in the midst of restoring an International 1568 tractor, a rare model with a factory-issue V-8 engine. He had bought it at a sale thinking they could modify it for tractor pulls, but then decided it would be more valuable restored to original condition. Zach was finishing work on a three-spike hay bale carrier built for a neighbor. Mounted on the front of a tractor high loader, it looked more like a medieval weapon of war than a farm implement but was unlikely to menace anything more dangerous than a large bale of hay. His welding arc flickered, hissing and humming and filling the shop with a blue haze of wispy, acrid smoke. In the back of the shop, a radio covered in years of dust and grime sat on a long workbench. It was hot-wired to the light switch, its dial tuned to a country music station and the volume low. Zach's welder obscured the singer's voice. Satisfied with his welds, Zach hefted a portable grinder to smooth some rough edges. The grinder sent jets of fiery orange sparks splaying across the floor, some deflecting off one of the 1568 tractor's dual tires and bouncing to the floor before going cold. Boy the shop cat, crossing to the rear of the building, made a wide detour around the sparks. While Zach finished the bale spike and spray-painted it red, Del inspected some dings in the 1568's sheet metal and applied body filler, scraping it smooth. A pungent acetone and organic odor permeated the shop, already gauzy blue from smoke, sparks, dust, and colloidal suspensions of metal and paint. The shop was like this most winter days.

11

March 10
Flood

A gentle rain started in the night. By mid-morning, it was pummeling the spare landscape. Puddles in roads swelled to pools and mirrored an angry sky filled with dark-bellied clouds. Ditches overflowed, streams became rivers, and fields became lakes. Zach looked at the rising floodwaters in the stream below his house and reached for his phone.

"Dad, I need your help." His voice betrayed little emotion. "Creek's flooding. Water's already over the road and rising. There's a cow and her newborn calf marooned on a little peninsula of land in a creek bend."

"They still there?" Del asked.

"I can see them right now. I think I can save the calf before it gets in the creek, but I need your help to do it."

"You think we can get to them?" Del asked, unsure what Zach had in mind.

"It's worth a try, but we gotta hurry. If she crosses the creek, the calf will follow. Then it's a gonner for sure. It'll get swept away."

Del frowned as he grabbed his rain slicker. The porch door slammed behind him as he crossed the rain-splattered driveway to his truck. Approaching Zach's house minutes later, he saw the floodwater. A creek crossed the road just beyond where the lane entrance led uphill to the house. The small cement bridge banisters were already partly submerged. Water covered the road for a hundred yards or more in both directions, and the lower end of the lane was under water. Menacing whirlpools tugged at fences and trees. Del had driven this road a thousand times, knew where the turn should be. All he saw now was a sea of angry brown water. A swift current carried branches and debris across the road. Some of the flotsam was lodging against fences, creating little dams and eddies. Del hesitated, then shifted into four-wheel drive.

"I'll go slow," he muttered to himself.

He unconsciously gripped the steering wheel a little too hard as he nudged the truck forward. Everything was submerged beneath the rapidly moving water. Even the bases of the fence posts and the mailbox were underwater. Del fought the tendency to drift or steer in the direction of the moving water. He could sense the hypnotic movement of the water playing tricks on his mind and could see how people in floods got in trouble. With neither ditch visible, he concentrated on steering the truck straight ahead, then made a hard left turn just past the mailbox and where the entrance road would be. "Road ought to be about here," he said to himself, grimacing. "Reckon nobody's moved it since I was here last."

Zach, wearing a short green slicker, waited in the rain outside his house. Water dripped in a stream from the bill of his grease-stained cap.

"Water ain't over the bridge banisters yet," Del said, grinning and rolling down the truck window as he stopped next to Zach. In the open window water drops splashed on his sleeve. "Where's the cow and calf?"

"Down there," Zach motioned off to the north. "I got a ladder ready. If we throw it across the creek, I can crawl over on it and get the calf. Creek's pretty narrow there. Can't get to it from the other side."

"Sounds a little scary. You think the cow can swim across in this?"

"She'll make it. Plenty of bends in the creek where she'll get pushed against a bank if she gets in trouble. Creek's not all that big anyway. Calf's what worries me."

The creek was overflowing its banks except in a couple of slightly elevated areas. One was where the cow and calf were marooned. Del backed up as close to the creek as he dared. Water was just ready to overflow into the grass where he stopped. They raised the ladder, holding it with a rope tied near the top, and lowered it out across the narrow creek toward the opposite bank. Once in place, it formed a precarious bridge to the slightly higher ground on the opposite bank. Del stood on the near end of the ladder to hold it steady.

"Ladder's practically in the water. You sure you still want to do this?" he questioned.

"Well, not particularly, but it beats swimming over."

Zach crouched low, held onto the sides of the ladder with his hands for balance, and warily crept across the ladder to the opposite bank. It was scarier than he thought. The little black calf was new, maybe only a day old. Lying wet and shivering on the cold, rain-soaked ground beside its mother, it looked helpless and uncomprehending. Zach could see that glassy-eyed look that usually signaled an animal was weak, maybe near exhaustion from the cold, wet rain. Without hesitating, he leaped off the ladder, reached down, and scooped up the slippery calf and cradled it in his arms. A flood of water drained off its back. He backed up a few steps, half-turned, and crouched to step on the ladder, and then looked down at the fast-moving water right at his feet. Realizing he was in trouble, he let out a gasp. This was going to be more difficult than he'd imagined. Now he had only one hand free to hold the edge of the ladder. He glanced back at the confused cow. She was just beginning to comprehend that her calf had been abducted. Zach knew she might charge. If only he could tell her what he was doing. His heart pounded. Whatever he did it would have to be quick. The muddy, swollen creek was almost touching the ladder. If his foot slipped, he and the calf would both be in the creek. For a fleeting moment he wasn't so sure this had been a good idea. Should he just leave the calf? Del saw the cow shake her head and lower it. Her black eyes stared, glistening.

"Watch out! I think she's gonna fight ya!" he yelled.

Zach reacted. Didn't think. He shifted the calf slightly, tucking it under his left arm so he could hold it firmly behind its front legs. With the calf's back legs dangling and rain pelting his face, he crouched low and stepped onto the horizontal ladder, using his free hand to grasp the rungs for balance. Grimacing, he lurched across to the opposite shore, frantically trying to maintain his balance and his grip on the calf. The cow swung her head and charged, but Zach was already gone. She stopped short, catching a hoof on the edge of the ladder just as Zach stumbled and leaped off the opposite side. The cow's foot caused the ladder to jerk, surprising Del. He yanked hard, pulling the ladder free of the cow's hoof, but the far end fell into the fast-moving water and was instantly swept sideways downstream. For an instant Del thought he'd lose his grip. He had visions of a thirty-foot aluminum ladder careening downstream. He pulled hard, slipped in the wet grass, and fell backward on one knee, but he maintained his grip and dragged the ladder out of the creek. With rain in his face, he scrambled backward to get upright, then hefted the ladder onto the flat bed of the truck just as the cow plunged into the swirling creek.

"Get the calf up on the back of the truck!" he yelled to Zach.

He wanted the cow to see her calf. Cow and calf bonds were strong, and they suspected she would try to cross. Almost as soon as the cow stepped off the bank, she was swept fifteen or twenty yards downstream, but her forward momentum allowed her to reach the opposite bank where the creek bent sharply back to the right. The current pushed her against the bank just as Zach thought might happen. She struggled, breath steamy and convulsive, hooves digging at the soft bank. Del and Zach saw her gain her footing in the swirling water and knew her maternal instincts were strong. But she needed more than instincts. Maybe the current was too swift? She pushed hard. With back arched and head low, she lurched upward and attempted to heave her large body out of the water, then slipped and skidded backward.

"Oh-oh! There she goes," Zach said in disbelief, suddenly feeling helpless.

But the cow gamely bobbed up again. This time, with help from a little eddy current, she heaved her dripping body up the bank in a desperate burst of energy, water pouring off her back in sheets. With tail held high, she splashed free of the stream and stood for a moment as if attempting to comprehend what had just happened, then looked toward the truck. She saw her calf. Now she would follow them anywhere.

It had been more than a week since the flood. Morning chores—feeding cattle, checking on expecting cows, looking after newborns, which were coming at the rate of two, three, sometimes four a day—seemed almost routine again. After the flood the weather warmed dramatically and brave spring grasses flushed pastures with green. Prematurely green, perhaps, but cattle abandoned hay en masse for the rank essence of new grass. It would be another month before there would be sufficient grass to sustain the herds. But new grass was a powerful elixir, and after months of dry hay, the purgative effects of fresh grass on bovine intestines, even in small amounts, was dramatic. Worse still, many cows were beginning to shed their winter coats, their long, rough, winter hair falling out in unsightly patches in a process that would continue for a month. Del said the cattle never looked worse than during the first few weeks on spring grass. If the cows seemed harrowed and ungroomed, their new calves were, paradoxically, the opposite—fresh and winsome with thick, woolly coats, big dark eyes, floppy ears, and tails that circled round and round when they frolicked. Their legs were still far too long, and they bucked and kicked and chased after each other in larkish abandon beneath spring skies.

Sod pastures were solid enough to drive across now, not squishy and soft like a week or more ago. Del hated to make deep ruts in pastures, but sometimes it was necessary to get hay to the cattle. During wet periods, he tried to feed close to fences or roads where access was easy. This morning Zach returned an errant cow that had gotten over an electric fence and wandered into an adjacent pasture. In another pasture an impertinent calf was entangled in a brush pile. Carefully picking his way into the brush, Zach grabbed the frightened calf and freed it by physically lifting it over a blockade of branches. He wondered how it had gotten in there in the first place. While driving alongside a stream after feeding the last cowherd, he saw a turkey vulture flush from a partly concealed position in the streambed. The vulture was a recently returned migrant, and seeing it flush from such an unlikely place piqued his interest.

"Might be a dead calf," he muttered, sliding out of the truck seat and peering over the bank at the stream below. His suspicions were confirmed when he saw the body of a little Black Angus calf on a shrubby island of stream gravel. Its limp body was wrapped around a sapling like a discarded rug.

Poor thing probably got separated from its mother in the flood, or was trying to follow her, he thought. If the vulture hadn't flushed, he wouldn't have noticed it. He'd already been by this spot a couple of times since the flood. He wasn't aware of a cow missing a calf, and couldn't dismiss the possibility that the calf belonged to a neighbor. It could easily have been swept downstream this far. Zach made a mental note to talk to the neighbor later. He'd also check their own herd to see if he could spot a cow with a swollen udder that looked like she hadn't been nursed.

With a small rope from the truck, Zach climbed down the bank and jumped across to the gravel bar where the carcass lay. He slipped a noose around its two back legs, dragged the lifeless calf up the bank, and tied it to an arm of the bale loader on the back of the truck. It looked small. Its onceshiny black coat was matted and mud-stained. The skin on its face was soft and peeling in whitish patches, perhaps where the vulture had pulled at it, but its eyes, dull and clouded gray, weren't punctured. A small section of its soft underbelly had been severed, and a length of cold, ash-gray intestine, coiled and kinked, spilled out like old wet rope. Zach threw the carcass onto a large brush pile that would be burned in a few weeks.

When Zach told Del about the calf, Del didn't seem surprised. "That was one of the worst winter floods we've had in a long time," Del said. "We coulda lost more."

One morning after the flood, Del and Caitlin were interrupted when a neighbor's dog ran into a pasture by the house and began chasing the cattle. Caitlin recognized the dog and called the owner, a neighbor, who lived about two miles away, but she wasn't satisfied that much would be resolved.

"When I yelled at that dog, it bolted fast as it could run," Del said. "I bet it's been shot at already for chasing cows." After a pause, he added, "It may get shot at again."

Caitlin knew what he meant. "We've got enough trouble with our own dog. We don't need somebody else's dog spooking our cattle."

They had hoped their dog could be trained to help with cattle roundups, but it had showed little aptitude for working cattle so far. A few days ago, when Caitlin was playing with her, she jumped up, ran off into the pasture, and started barking at the heifers. They spooked, ran right through a fence, tearing it down, and continued in a panic across the blacktop road north of the house and a half mile up an adjoining gravel road before they finally stopped, exhausted, all the while with the crazed dog nipping at their heels and barking furiously.

"At least there weren't any cars on the road. Imagine if a truck or something was coming," Caitlin said. "I'm just glad you and Zach were around to get them all back."

"Me too," Del said, laughing, but he'd been furious with the dog and had kept her tied up most of the time since then. "That dog training video you bought ain't no good at all. She shows every one of the characteristics they say is typical of an untrained dog, only I'd say an untrainable dog."

Toward the end of the month, the weather turned cold again. Everybody decided spring hadn't arrived quite yet and dragged out insulated coveralls to ward off the morning chill in the shop. The red International 1680 rotary combine parked in the shop was in the midst of its annual winter maintenance. It was the last of the three combines to get serviced. Inside the shop the combine seemed even more enormous than it did outside, filling the entire front half of the shop, its grain tank extenders just inches below the rafters. A rear vacuum cover gasket wasn't functioning properly, which let the engine blow a little oil into the cylinders. With nearly 6000 hours on the engine, it was due for an engine overhaul. Del had spent a thousand dollars on a replacement kit, which included new sleeves, pistons, and connecting rods. A year ago a sleeve in the cylinder of another combine engine developed a tiny hole. It could have been just the tiniest spot, but the high pressure inside the diesel cylinder blew fuel and air through the hole in the sleeve and into the engine's cooling system, and the pressure exploded the radiator.

Del laughed thinking about what happened. "Radiators weren't built to withstand that kind of pressure. When it blew, it sprayed steam and scalding water everywhere. Didn't burn nobody though."

Zach asked him what he thought would cause the hole in the first place.

"Hard to say. Maybe a speck of dirt or something got stuck on the outside wall of a sleeve. That might alter the cooling and start burning the sleeve. If a tiny rusty spot developed there it would weaken the sleeve and maybe a hole would develop."

"That why you always got me changing coolant in something?" Harlan teased, wiping a greasy hand across his brow.

"You know we change coolant in every engine on a three-year schedule. That's the only problem we've ever had."

Del carefully washed the anti-rust fluid from the sleeves and pistons and tapped each sleeve into place in the engine block, which was sitting on its side on wooden blocks on the floor. He then lubricated the cylinder's surface and seated two rubber compression rings around the middle of each piston, and below them inserted a spring-loaded oil compression ring. Then he fastened a connecting rod to the bottom of each piston with another spring-loaded safety clamp and placed two inserts around the cupped bottom end of each connecting rod where it would clamp around the crankshaft. Like completing a three-dimensional puzzle, he eased each piston and connecting rod into place in the engine block, turning the crankshaft slightly so he could couple each piston and rod to the crankshaft. With everything in place, he torqued the bearing and rod bolts to exact tightness specifications, checking the torque on each bolt twice. By noon he was ready to add a new head gasket and reassemble the head of the engine.

Zach changed a sprocket that moved a sieve on the inside of the combine, then turned his attention to aligning the rotary cylinder, which was the business end of threshing in the combine. In most combines, the threshing cylinder spins perpendicular to the flow of incoming grain, but in rotary combines like this one the cylinder revolves lengthwise to the inflow of grain, in theory allowing more grain and straw inflows without clogging. Rotaries didn't require much maintenance but they weren't maintenance free. Zach had just finished changing all of the curved threshing bars on the rotary cylinder. "I replaced every cylinder bar," he told Harlan. "I think I skinned my knuckles on nearly all of them."

Zach adjusted the cylinder spacing to align it within the threshing cage. If the spacing wasn't the same throughout, the cylinder would crack grain at one end and let it pass through unharvested at the other. Unable to get the spacing to match exactly, he phoned the dealer. A mechanic suggested reversing a jam nut. It didn't help. He looked at two small brackets a previous owner had welded onto the sides of the cylinder cage. It wasn't obvious why they had been added, but they could be the problem. He cut both of them off with a torch. That was when he heard Harlan and Del discussing dinner. He was ready for a break.

12

April 1
Spring

April. Bobwhite quail begin calling; farmers spread anhydrous ammonia and fertilizer on crop ground; farm machinery readied as farmers begin preparing fields for spring planting; corn crop usually planted by middle of month if weather permits; calves vaccinated and castrated during annual spring roundup; cattle moved to spring pastures; family garden ploughed and seed bed prepared; honeybees swarm; crappie spawn by end of month.

A gusty wind buffeted the countryside. Pastures wore newly minted, close-cropped carpets of green, and fallow fields were painted with twenty-acre-long brushstrokes of soft pink. The swatches of color were henbit, a diminutive plant in the mint family, and an alien from Europe. Caitlin called it chicken peck. It usually occurred in lawns or roadsides and waste areas. This spring it blanketed fallow fields—even corn stubble treated the previous year with atrazine, and soybean fields planted with Roundup Ready beans. Millions of its tiny blossoms brightened the landscape with pastel color. How could one tiny plant have gotten so numerous? Del thought it probably got mixed in with a batch of last year's bean or corn seed. No one showed much concern. Across the county preparations for spring planting were underway. Corn was the immediate focus.

Spring touched the evolving landscape in many ways. Purple martins, recently returned from South American wintering grounds, chatted amicably and inspected a bird apartment complex atop a long pole behind the Montgomery house. Caitlin was fond of the martins. She liked their noisy exuberance and had read that they ate lots of mosquitoes. An eastern phoebe sang its name from a small cattle shed where a calf had been born earlier in the winter. Now the building sheltered the phoebe's nest and eggs.

Overhead a great blue heron, gaunt and ungainly, labored in flight, pressing northward, neck folded, long legs trailing behind. The wind repeatedly buffeted it sideways, redoubling the struggling creature's efforts to stay aloft and its determination to heed migratory instincts.

Del left the farmhouse early to check his cattle. Over the last few days with Zach and Harlan helping, he had moved herds to summer pastures. Moving cattle was a twice-a-year event, one that mirrored the celestial rhythm of the sun's march across the equator, ushering in seasons of plenty and seasons of want. The rhythm of cattle moving was easier in the fall. At that time, with grass diminishing, cattle were hungry and eager to move. In the spring, amidst a flush of new grass, they loitered or scattered, lost interest in hay, and become indifferent to roundups and men who organized them. Spring roundups also were complicated by calves with more energy than discipline. They acted frisky, romped away, and distracted their mothers whose maternal instincts overcame herding urges. Del complained that he might as well be herding stray cats or shooing chickens. The three of them spent most of the previous morning corralling and loading one herd to move to summer pasture.

Fewer new calves were arriving now, no more than one or two a day. It was a dramatic reduction from February and early March. There was none this morning. The long winter feeding and calving season was ending. By now, farmers had spring planting on their mind. Planting required herculean efforts for a few days or weeks, but then the work, except for an application of herbicide, was mostly over until harvest. After that only the watching and the worrying about the weather remained. Cattle also required less attention now than during the winter months, and they pretty much took care of themselves through the spring and summer.

Driving slowly through each herd, Del's eyes roamed over cows and calves, appraising their condition with a skill that came with years of familiarity. Where urban dwellers saw milling cows in a pasture, he saw the thriftiness of animals in good condition, or little problems like subtle changes in gaits, animals without appetites, weepy eyes, stiff limbs, sore feet, bloated bellies, scours, pregnancy, eminent births, nursing cows, and old age. Above all he saw quality and bloodlines in cattle. Every cattleman appreciated good-looking livestock. They knew it when they saw it. His livestock, any livestock, was almost an open book to be read.

One calf had weepy eyes, something requiring a follow-up. Another cow and calf had a problem he couldn't diagnose for sure—the calf, more than two weeks old, was suckling but looked gaunt and hungry. Maybe the cow didn't have enough milk. He separated the cow and calf and herded them into a corral.

Harlan arrived about eight, and Del helped him get started putting an electric fence across a fescue field to restrict a group of heifers. "I want to keep them off the east half of the field so we can harvest the seed in June."

"What about the creek?" Harlan inquired. "Can't get the wire low enough over the water."

"Use a couple small wooden panels. Suspend them from a heavy wire or cable below the electric one. If the creek comes up, the water will flow around the panels. Use the Gaucho charger. The solar charger's already in use."

"What battery you want to use?"

Del looked over a collection of seven or eight large car and truck batteries sitting on the floor at the front of the shop, some long and heavy, others smaller. "This one'll do," he said, pointing to a yellow one. "It's charged. Ready to go."

It was a day of errands. Running a farm sometimes was like that. Del still enjoyed operating machinery, but he'd done his time on a tractor seat. Somebody had to maintain the flow of supplies and, nostalgia or not, it kept him busy supplying Zach and Harlan with what they needed in the fields—herbicide, fertilizer, seed, fuel, replacement parts, moving equipment and trucks from field to field, operating machinery while they ate their meals so time wouldn't be wasted, hauling grain to town to sell. If keeping things operating smoothly meant that he didn't do that much fieldwork himself anymore, that was fine with him. Farming had changed. When he first started, even small family farms were thriving. Almost no one had a permanent hired hand, yet people made decent livings on their farms. Now it was harder. Profit margins were slimmer. Farmers had to play in an international commodities arena. It came down to volume and efficiency. A farmer had to do both to survive, or else work in a niche market like organics, or some specialty like growing cucumbers.

He still did plenty of hard work, still put in the long hours, but sometimes he wondered if it wasn't a little harder now to keep going for days, even weeks, during the long planting and harvest seasons. When he and Caitlin were just starting and couldn't afford a hired hand, she'd helped him. Now he wondered how she'd managed to help so much, especially when the kids were little. The two of them had done all the work themselves and with smaller, less-efficient farm equipment. And they'd still found time to have a little fun, too, and had managed a vacation almost every year.

Today farm equipment is larger and brings more capacity to a farm business. A farmer can take on much more acreage now, but it takes two or three people to keep high-capacity machinery fully employed during harvests and haying and, with more land to farm, everything is multiplied—more fences to maintain, more pastures to mow, more spraying, more fertilizer to spread, and more equipment maintenance. There were times during the year when Del pushed efficiency to the limit. He felt that with three men the farm operation was, at present, about as close to peak efficiency as possible, and he could foresee the need for another hired hand if he continued to expand.

By 9:20 in the morning, Del had checked cattle at four locations, briefly discussed the day's plans with Zach, who had been applying anhydrous ammonia on the Somerset farm since seven in the morning, helped Harlan organize electric fence supplies, paid a fertilizer bill, and deposited a check at the bank in Prairie Point. And the morning was just beginning. At the homeplace, he hooked a long, black cattle trailer to one of his dual-wheeled pickups and picked up Lewis Larousse in Prairie Point. The two of them headed to the United Producers auction about forty-five minutes away. Lewis's brother had bought some bred cows last night at the sale, then called Del about 11 p.m. and asked if he had time to pick them up today. Said he had to work and couldn't get them. It was late and an inconvenience, and Del had a full schedule planned, but he didn't hesitate to help out. Someday he might need a favor.

Lewis was ruggedly built with a graying beard and a little younger than Del. He lived in Prairie Point, but his brother lived a few miles out of town on a farm between new Amish neighbors. Del admired the Amish work ethic and family commitments, and he knew they'd never be on welfare, but their rejection of modern farming equipment seemed rift with contradiction. Farming was hard enough work with modern equipment, he thought.

He couldn't imagine farming the old way. Driving south they passed an Amish homestead where a barn raising was in progress. Fifteen or twenty buggies were lined up outside a wire fence by the road. Inside the fence, in the pasture, there was an equal number of sleek, leggy, dark-coated horses. The barn, a sizable two-story, gambrel-roofed structure, was studded and framed like a giant erector set. Men in dark bib overalls, blue shirts, and black hats swarmed over the building like color-coded ants with hammers and saws. A steady stream of board siding and roofing materials was being hoisted up from below and nailed into position. A slender teenage girl in a bonnet and long, pale blue dress stood in the yard motioning to someone. Several young boys in little bib overalls, miniatures of their fathers, stood near a wagon loaded with lumber.

"My nephew has gotten to know several of them pretty well," Del said. "They come into his grocery store and buy a lot of food. Not all basic stuff like you might think. They get chicken and beef, and the young ones get candy and snacks, even ice cream. They're good workers, and we've hired them several times to work on barns."

"Well, my brother doesn't like them so much," Lewis answered.

Del's eyebrows arched. "Oh, how's that?"

"My brother and Jeb Gannon have trouble with them trespassing all the time. You know, they walk everywhere."

"Kids?" Del questioned.

"No! Not kids. Big ones. Come walking right through the pastures, cutting through the cattle," Lewis retorted, his face tightening. "Gannon stopped a couple of them a few days ago. Told them to get off his property. Said he didn't know what they called it back in Tennessee where they came from, but here in Missouri we call it trespassin.'"

"Did they leave?"

"Oh, yeah. Ain't been back neither."

Del decided to let the conversation drop, then turned up the lane toward Lewis's brother's place. "Just pull in there," Lewis said, pointing toward a gate. "Circle around back. We can unload the cows by the barn."

On the way home, Del called the vet about the cow with the sickly calf. Dr. Barkley wasn't around, so he called another vet who said he would be at his office for about an hour, if Del could just load them in a trailer and bring them by.

Within fifteen minutes, cow and calf were loaded and en route to Ashton. "She's got mastitis," the vet said, feeling the cow's udder. "Spread all through it. Here, feel how hard it is. She doesn't have a drop of milk. How old's the calf?"

Del paused, thinking, "Nearly three weeks, I guess."

"Little feller must be stealing milk from another cow," the vet responded, "otherwise he'd been a goner by now. He's not getting enough or he wouldn't be so gaunt looking."

"I'll take the cow to the sale barn Thursday," Del said with resignation. "Maybe keep the calf. One more to add to the orphan nursery. Being so puny he won't bring much at a sale, but they're hard to feed at this age."

"I know what you're saying. Too young for solid food but hard to teach to take formula from a bucket."

On the way home, Del stopped at A&C's Machine shop at the edge of town. Two bearings on the pitman shaft that powered the sickle bar on the swather they bought last November needed replacing, and the shop was having difficulty identifying the part by number. Leroy, who worked at the parts counter, was gruff and direct. He stared at the computer screen, mouse-clicking through an endless succession of exploded parts diagrams. After a few minutes, he pushed a dirty AGCO cap further back on his balding head and said, "I can't be sure of which damn bearings and trace to order until I get an exact measurement from you of the size of the bearing on the shaft. There's been so many mergers in the farm machinery business that tracking down these part numbers is getting near impossible."

Del figured Leroy was right. He knew that AGCO alone consisted of at least five or six former machinery companies, among them Allis-Chalmers, Case, White, Hesston, New Idea, and Massey Ferguson, and most of them were themselves composed of earlier mergers. Removing the bearings would be difficult; he might have to cut them out with a torch, but Del figured he'd have to do it eventually anyway. He left, telling Leroy he'd call him with the measurements.

Caitlin finished her midday meal and was ready to return to work at the post office when the men arrived. Her part-time job had become almost full time lately. For the past week she'd worked every day. Del could tell. They'd been eating lots of cold cuts. Today it was cold cheese sandwiches, potato chips, and ice tea. At dinner Harlan said the clutch had gone out on the old dodge truck while he was putting up the electric fence. He'd gotten back to the homeplace by starting the truck in gear, then shutting off the engine when he needed to stop.

"I knew that was about to happen," Del said. "Clutch has been jumping and grinding for several days."

Del finished his dinner with a giant salad—half a head of lettuce unfastidiously cut up in a large bowl—then turned to Zach. "How you getting along applying the ammonium hydrate?"

"Soil's almost perfect. Good moisture." Then a wide grin spread across Zach's face. "Somebody got into one of the anhydrous tanks last night."

"Stole some again?"

"Yeah, meth lab guys, I reckon. Seems like whenever they're making drugs they think they can just help themselves to some of our anhydrous ammonia."

"They leave some of their junk around?"

"Left part of an old bicycle tire inner tube on the ground. Cut the ends of the fingers off your safety gloves again. They were in the toolbox on the wagon. You got any idea why they do that?"

"Maybe so they can handle that stuff better in the dark. I don't know. Even with the fingertips cut off, the gloves would still protect them some. But it would freeze their fingers off if they spilled it."

"I could see where they bled off some. There was a patch of dead grass by the tank."

"They shut off the valve?"

"Looked like they did. Probably didn't take much."

Del shrugged. "Nothing much we can do. You can't hide those big white tanks. Easy to see. Too dangerous to put them in a shed."

"You want to call the sheriff?"

"Not worth the bother. Those guys usually don't take much. Probably won't be back for a while, anyway, so wouldn't do no good for the deputy to do a stakeout. Wish they'd quit ruining my safety gloves though. I'm gettin' tired of buying gloves for them to cut up."

"I'll need more anhydrous by mid-afternoon," Zach reminded Del. "I'm almost finished on the Somerset place. Should be moving to the Hyatt farm by evening."

"I'll get more from the Farmer's Exchange in Grand Forks," Del said.

"You'd think some of those meth guys would get hurt or killed stealing that stuff."

"Probably have. We just haven't heard about it. Get a leak in a hose and you can't get anywhere near it until the wind clears the gas away."

Anhydrous ammonia was mostly just nitrogen stored under high pressure in big tanks. Farmers used it because it was an inexpensive way to add nitrogen to the soil. It was made largely from natural gas, steam, and air, but it was hard to produce in small quantities. You couldn't buy just a gallon or two. That was why the meth lab guys went after the anhydrous tanks that farmers used. Like white flags in the night, the tanks were easy targets. It was easy to bleed off the small amounts a meth cooker would need.

Farmers applied anhydrous ammonia by towing the tanks behind their field cultivators or planters. Even if the application required an extra trip over a field, the savings could be at least ten dollars an acre over regular fertilizer—a big savings to cash-starved farmers with hundreds of acres under cultivation. Farmers still applied fertilizer when planting, but only the potassium and phosphorus components. Anhydrous ammonia had been used in the Great Plains wheat belt long before it became popular eastward. That was partly because the drier and lighter soils of the Great Plains were more forgiving of tractors struggling to pull big chisel ploughs, which were required to knife in the anhydrous ammonia to a suitable depth. New and more powerful jumbo tractors of the last few decades, however, had made it possible to apply anhydrous ammonia even in the heavier and wetter soils of the Upper Midwest. Some guys even pulled harrows or planters along with the anhydrous tank and chisel. Hooked together, the string of equipment looked like a caravan moving slowly back and forth across fields.

Behind each chisel knife there was a small tube that fed the anhydrous ammonia directly into the soil. At normal temperatures it changed instantly from a liquid to a gas and combined with moisture in the soil where it would spread six inches or more laterally each way. That's how it all worked in theory.

In practice anhydrous ammonia was finicky stuff, dangerous to transport and to use. It boiled as a gas at -28 degrees Fahrenheit, a temperature so low that for agriculture it had to be pressurized in 1000-gallon nurse tanks to make it efficient to use. Getting the right amount of it applied to a field was no simple matter. It had to be applied just right—at constant speed, constant depth, and when the soil was not too wet or too dry—a Cinderella suite of conditions that didn't always fall into place. Flow rates were affected by ambient temperature, tank pressure, soil moisture, and even by something as simple as a plugged applicator tube, so operators had to pay attention to pressure gauges, adjust flow rates carefully, and drive at specific speeds. Even the slope of a field could affect how evenly it got applied. It corroded copper and brass fittings as well as nylon and polyethylene applicator hoses, and absorbed water so quickly that it would freeze skin instantly, causing severe dehydration burns. A drop or two in an eye could cause instant blindness. Farmers, or anyone using it, had to be licensed. Heavy rubber gloves, goggles, and respirators were standard safety equipment. Nurse tanks were painted white to reflect light and reduce the chance of a tank getting too hot and exploding.

Del said it was a wonder farmers or anybody else messed with the stuff. "Goes to show what people will do to save a little money."

Remarkably, in agriculture there had been very few accidents. The dark side of anhydrous ammonia, though, was its role, as a bit player, in methamphetamine production. It was a theater of absurdity—a fertilizer component that helped feed the world was being used in the production of an illegal substance that destroyed the lives of people who used it. Meth could be made without it, but using anhydrous ammonia was said to yield a better product. What bothered Del was that if the co-ops couldn't find a way to safeguard it, they might decide not to sell it because of liability issues and increased costs for insurance and clean ups. A few had already stopped selling it. In Grand Forks, where Del bought his anhydrous, thieves had stolen it so often the co-op had almost given up preventing thefts. A co-op employee told Del that one inept thief tried drilling through the side of a pressurized tank. Another left a valve open and drained an entire tank during the night—about eight hundred dollars' worth.

"I heard there's some new additives being tested," Zach said. "Then it's no good for meth."

"I hope that's true before somebody innocent gets hurt from these guys messing around in the middle of the night," Del said.

After dinner, Del examined the truck Harlan had been driving. The gearshift was locked. He'd get Harlan to work on the transmission. Harlan was good at fixing old trucks and cars. His GMC truck was testament to that. A few minutes later, Del was on the phone with Leroy at A&C's, giving him measurements for the bearings. The shaft measurement was 370/1000th, slightly less than what they estimated. He was glad Leroy told him to make the measurements before ordering the bearings.

Del had to get more anhydrous to Zach by late afternoon, and that meant a trip to Grand Forks, fifteen miles away. At 2:45 p.m., he stopped at the Farm Credit Services office. Fifteen minutes later, at the Missouri Farmers Exchange, he paid for a delivery of bulk diesel fuel. By 3:15 p.m., he was on his way home with two anhydrous ammonia tanks in tow. The wagon's running gears towed smooth and straight up to about thirty-five miles an hour, but any faster and there was a disconcerting tendency for them to weave. He could feel the weight of the heavy tanks tugging on his truck. He would take his time. He'd be getting two a day for the next several days.

Farmers were not planting corn yet. Soil temperatures were edging up close to fifty degrees, cool for corn, but that would change in a week if the weather held. Zach was waiting when Del arrived with the anhydrous tanks. It was 4:00 p.m., and Del figured he had enough time to start spreading a little bulk fertilizer—in this case just phosphorus and potassium. He switched trucks, drove his old Dodge truck with its white, V-sided hopper and chain-drive unloading mechanism to the fertilizer plant in Prairie Point. While the truck was being filled, he checked a fertilizer chart to see how to set the spread rate on the truck. The mix of fertilizer affected its weight, so the gate opening at the back of the truck had to be set to the proper weight of the fertilizer in order to get a full sixty-foot swath—too wide and it would throw the fertilizer too far, too narrow and it wouldn't throw far enough.

Back in the field thirty minutes later, the engine screamed as the old truck, a twin-axle with tandem duels and faded orange cab, lurched and bounced across the bare, clod-strewn field. Two horizontal metal spinners at the rear of the truck flung buckshot-sized pellets thirty feet to either side, pelting the lumpy soil like sleet. Halfway through the first pass, the engine lost rpm, sputtered, and died.

"Sounds like the fuel pump," Del muttered to himself. With the ignition switch still on, he jerked the fuel line free and checked the fuel flow. The pump whined and spurted fuel irregularly, sometimes stopping completely for several seconds before refilling the lines. He got the engine started again, but before he completed another pass through the field, it died again. He glanced at his watch, then pulled out his cellphone and called a parts dealer in Grand Forks. They had a fuel pump. There was enough time to pick it up late this evening, his second thirty-mile round-trip to Grand Forks today. With luck he could get the truck restarted and back to the shop before it quit again, get the new pump, replace it tonight, and be ready tomorrow morning. It would be dark soon. Chorus frogs were already calling in the damp ditches along the highway.

13

April 4
Winter Tractor Pull

The machinery was vintage, but the technology avant-garde. Smoke and an ear-splitting caterwaul filled the arena as engines screamed, giant tires dug furiously, front wheels reared, and fire belched from exhaust manifolds. It was the National Antique and Classic Tractor pull championship, the grand finale of their winter pulling season. Each gut-wrenching pull lasted barely ten seconds, and pull distances were laser measured to a tenth of an inch. At the end of the meet, accumulated points from the season would be added up and winners declared. Accolades were minimal—your name and picture published on a web site, maybe a trophy. It was mainly bragging rights. Top prizes were barely enough gas money to get to the meets. The men, and a few women, competed because they loved the tumultuous, high-voltage atmosphere of noise and earth and exhaust, a place where raw horsepower was the sovereign arbitrator.

Del and Zach were on the road before seven for the three-hour drive to the meet. Their route took them through ordered, mostly rolling farmland of central Missouri. Fields drifted past the truck windows as they talked, and their banter seldom strayed from farm topics. Zach wasn't a member of this pulling association, but he thought it would be fun to compete, and anybody with a tractor older than 1955 and a thirty-five-dollar entry fee could enter. His gleaming red Farmall 460, an early 1950s model with a big chrome exhaust and enough internal modifications to quadruple its original horsepower, was on the trailer behind their pickup.

Most tractor-pulling associations, including the one that Zach belonged to, scheduled meets during the summer and were regional in membership. Today's meet, by contrast, was a nationwide event with competitors coming from as far away as the Dakotas, Colorado, Texas, even New York. Zach and Del enjoyed the meets and the banter with tractor owners. Del said that nobody had a monopoly on good ideas, and he figured he could always learn something at these events.

It was an irony that the performances of these ancient Moline, Deere, International, Oliver, and Allis tractors—old models manufactured between about 1940 and 1955, and the antiques built prior to 1940—were being tweaked and measured with twenty-first-century technology. Some had been plucked from agricultural obscurity, abandoned and rusting in brushy fencerows. Others had been sold at farm sales decades ago. Restoration costs far exceeded their original selling prices of a few hundred dollars. In categories where extensive modifications were permitted, some of the old tractors boasted tens of thousands of dollars' worth of high-performance parts. These weren't their father's or grandfather's old farm tractors, even if they looked like the ones they used.

Tractor pulling had gained in popularity in recent decades but remained a niche sport, little noticed outside the mostly agricultural circles of those who participated. Its popularity was strongest in the Midwest farm belt where regional associations scheduled pulls nearly every weekend of the summer. More recently, it had gone international with pulling associations springing up in Europe, Australia, and even South Africa.

In the most basic pulling category, no modifications to the old tractors were permitted. Other categories allowed progressively more modifications as long as an original engine block was used and the engine burned some kind of gasoline or diesel. Unlike most summer pulls, this one was indoors, inside a sprawling coliseum-like building with dirt floors and large sliding doors on the sides. Two pulling lanes were marked off, permitting side-by-side events with scarcely a break in the action. Even with two lanes, there were so many competitors that pulls would continue throughout the day and into the evening. Eight or ten tractors and drivers, each waiting their moment on center stage, lined up on both sides of the arena. At the weigh-in, last-minute adjustments to total weight, weight distribution, carburetor performance, tire pressure, and other functions were tweaked. Little breezes kicked up dust around the scales. A sweet, butane-like smell of high-octane fuel permeated the area.

Onlookers watched from low bleachers, lounged against a slack cable fence encircling the pulling arena, or huddled in groups as they laughed and joked and listened to names and pull distances over an excessively loud

PA system. Tractor weight classes, in 250-pound increments, ranged from 3500- to an over-7500-pound category, and there were five performance divisions in each weight class. It added up to an opportunity for almost anyone to get a few seconds in the spotlight.

Three small maintenance tractors with leveling and packing implements circled the two clay tracks, grooming and conditioning the earth after each pull. They worked amidst befuddling noise and exhaust-laden air, but fortunately errant breezes wafted through the large, open driveway doors on each side of the building, bringing much-needed fresh air.

Registration and a food concessionary were located at the end of the stadium closest to the starting line for the pulls. The area was partly enclosed but not insulated from the noise and exhaust fumes. Attendees at the meet seemed largely oblivious to high-decibel noise so loud it made conversation difficult, and air so laden with exhaust fumes it might have been lethal to canaries, as they sat and ate around open tables next to the pulling arena. Most of these men had worked around noisy machinery all their lives. Many suffered hearing loss, but the noise and exhaust fumes were part of the show—a rock concert by another name—and they wouldn't have missed a minute of it.

Del and Zach spent much of the morning outside, examining tractors and talking with owners if they were around. The men, in caps, denim jeans and work boots, were mostly from rural backgrounds, but one man from Colorado, Zach discovered, owned several car washes, an unusual background for someone in this sport. A shiny yellow Minneapolis-Moline tractor with eye-popping flames painted across the engine cowling caught the interest of Del and Zach. They wanted to talk to the owner but couldn't find him.

Late in the morning, Zach registered for his pull. Del remembered when he had allowed Zach to compete in his first tractor pull. He was only nine years old then, but already experienced around tractors and machinery. On the farm he was driving tractors and helping in the fields by the time he was seven. Del had encouraged Zach's interest in mechanics and pulling during those early years, and had watched with pride as he gained confidence and knowledge.

In the 1970s, the sport of tractor pulling was still comparatively new, having evolved as an inevitable outgrowth of horse pulling contests, which were a staple of county and state fairs until the middle of the twentieth century. When Zach was a kid, most pulls were local and dominated by antique and classic division tractors. Competitors were likely to be your neighbors. Sleds used in the pulls were often just large, flat sheets of steel dragged down a dirt track. To add weight, men stepped onto the sled as it slowly passed by. Eventually, when enough men stepped on the sled, the increased weight stopped the tractor, a technique OSHA surely would have frowned upon. At that time, pulling speeds were slow and risks minimal, but the sport changed. Horsepower and pulling speeds increased. In high-powered divisions, tractors belching fire and rearing in front yanked fifty-thousand-pound sleds down a hundred-yard track at speeds of nearly twenty-five miles an hour before spinning wheels brought them to a halt. Pulling sleds had evolved too. New ones had a huge sliding weight that moved forward quickly to increase sled friction and sometimes spikes to further increase drag. The sleds were self-powered, too, so they could be quickly driven backward to the starting line after each pull.

Among tractor pullers competing in categories that allowed modifications, there was an inordinate preoccupation with reducing tractor weight. It was an obsession rivaling that of wrestlers and boxers who starve prior to weigh-ins to qualify for lower weight classes. The rational was similar—maximum power, minimum weight. It was not uncommon for tractors to pull in a weight class a thousand pounds or more below the original manufacturer's weight of the machine, while mustering two to ten times the original horsepower—a remarkable feat considering that original engine blocks had to be used in these pulls. There seemed to be almost no limit to what owners would do to reduce tractor weight—axles were downsized; cast iron wheel rims replaced with aluminum; front wheels miniaturized, some as thin as motorcycle tires; steering assemblages reduced; gas tanks replaced with tiny one- or two-gallon tanks mounted out front to serve double-duty as front end ballast; brake covers discarded; original brakes themselves replaced by newer, lighter disc brakes; fenders, drawbars, non-essential gears in transmissions jettisoned; and safety covers and fairings replaced with aluminum ones or abandoned altogether. Even wheelie bars, the mandatory braces behind the rear wheels to prevent tractors from flipping over backward, were minimized. Nothing escaped scrutiny. These were pumped-up tractors on SlimFast diets.

Pullers loved to invent names for their tractors, some whimsical, others intended to intimidate. Zach read some of the names on the tractors parked outside: Renegade, Survivor, Five-Star General, Tuff-N-Nuff, Bitin' Sow, The Undertaker, Black Magic, Two Extreme, Whoop Ass, Raging Bull, Chosen One. Others were more whimsical—Slow-Walkin' John, Ann's Big Crank, Temper Tantrum, Iowa Corn Burner, Little Miss Dynamite, Pocket Change—but the message was the same. Everybody did a little muscle flexing and chest pounding. Scrap Man, on the other hand, employed the opposite psychological tactic. A rusty old Oliver tractor from the late 1940s, it looked as if it had just been wrenched from somebody's brushy fencerow, but its rusted exterior belied numerous modifications hidden inside.

Judges, of course, were free to challenge anything suspicious, and they often did. An unusual performance could prompt an inspection, even a teardown that involved partly disassembling an engine—at the owner's expense. Penalties for infractions would be levied if irregularities were found.

Zach backed up to the pulling sled and two men connected his tractor to it. He knew he wouldn't win the heat. Two tractors in this weight division were former national champions, both 460s much like his own, but boasting transmissions, differentials, and engines with more performance features and power than his tractor. These were tractor-pulling aristocracy, agrarian glitterati with huge shaved tires and an attitude.

Del grinned as he looked at the two tractors waiting in line behind Zach, and commented to a stranger he'd been talking to, "I guess that's what a 460 looks like when it grows up."

The stranger replied, "Yep, but most of 'em that grow up that big usually blow up pretty quick too."

Earlier, Zach had eyed the bell housings, which contained the flywheel, clutch, and pressure plate on those two tractors. Both had scatter blankets, essentially bulletproof flak jackets of Kevlar securely covering the area where these parts were located. Because of these, their turbocharged engines would be allowed to run several hundred rpm faster than Zach's tractor. There had been accidents in the past, and the ruling was as much to protect bystanders as drivers from flying shrapnel if an explosion occurred. Such extreme measures hinted at the formidable horsepower these engines developed.

Zach's tractor wasn't turbocharged, but sometimes that worked to his advantage. When a turbocharged engine lost rpm under a heavy load, the turbocharger shut down and the tractor then pretty much stopped dead in its tracks.

Zach's naturally aspirated engine had less horsepower but more torque at low rpm than turbocharged engines and would continue to pull even as engine rpm slowed under the increasingly heavy sled weight. But he was outclassed today and barely managed to hang on to third place ahead of the shiny yellow Minneapolis-Moline with the flames and chrome he'd seen outside. The pulling distances of the two former national champions were within two inches of each other but sixteen feet further than Zach's.

After the pull, as Zach chained his tractor to the small flatbed trailer for the trip home, Del said, "You should have transferred more weight to the front."

"Yeah, I know," Zach responded, frowning. "Tractor's front end was up for most of the pull." Then, after a pause, he added, "Clay track was stickier than the dirt tracks I usually pull on. Harder to pull cause the tires wouldn't spin as much."

"Without a practice run it's hard to know that. But it's the same for everybody."

"I probably could have pulled a gear faster, too, high side of fourth, then torqued down to low fourth partway through. That might have helped, but there was no way I was going to beat those two guys."

Their post-tournament debriefing recapped what both men already knew. They'd attended and pulled in hundreds of meets, and on all kinds of earth and clay surfaces. They knew how the game was played and the limits of their tractors. Sometimes the track surface did make a difference but only so much. In the end, the tractor could do no more. That was why Zach had started building a new tractor, a from-the-ground-up model of his own design. Next year, he hoped to push performance limits well beyond what he could do now. But he was behind schedule. It was still an unpainted hunk of steel propped up at the back of the shop. Only the diesel engine and front wheel assembly were in place. Transmission, drive shaft, seat, fenders, tires, fuel tank, drawbar, and dozens of other parts were unassembled or still in boxes.

On the drive back to the farm, Del and Zach again studied the rolling central Missouri countryside. A couple of farmers were putting down anhydrous ammonia fertilizer along I-70, but nobody was planting corn. Winter's grip on the countryside was loosening, but capitulation came in fits and starts and would take time. Serviceberry shrubs, so named because they were supposed to have bloomed about the time circuit-riding preachers passed through to deliver their first service of the season, sprinkled bare woodlands with patches of snowy white. The mauve blush of redbud trees, another harbinger of spring, brightened ravines and hinted that planting season was near.

14

April 12
Coming of Age

The little black calf bucked and squirmed. Harlan grabbed its tail with one hand and the underside of its head with the other and pushed hard to guide the stubborn animal, perversely stiff-legged, forward into the cattle chute. Del yanked a lever, and the heavy iron gate of the squeeze chute rattled shut with a finality that didn't reflect the calf's continued struggles. Built for larger animals, the chute didn't fully immobilize the calf. To keep it still, Harlan placed a knee against its rear and pulled its tail upward, holding the calf in a kind of bovine hammerlock.

"You're just lucky that ain't a cow!" Del yelled.

"I sure as hell wouldn't be here inside this chute if this was a cow," Harlan grinned, a slightly bent cigarette dangling precariously from his lips as a plume of blue smoke rose above him.

Harlan had pretty much perfected the art of smoking without using his hands, perhaps out of necessity because most farm work required the use of both hands. He could run, corner, and catch a calf, and haul it into the squeeze chute, all the while keeping his lit cigarette under control. "This one's a bull!" he yelled out of the side of his mouth.

Del reached for a pliers-like device lying on the bed of the pickup truck parked next to the cattle chute. He placed a small green rubber ring, like a short, thick rubber band, over the extended upright tips of the device. With Harlan holding the calf firmly so it couldn't kick, Del reached between its legs, squeezed the pliers to expand the rubber ring, and slipped it over the calf's scrotum, making sure it was seated above its testicles. He released the pliers, and the rubber band held snug. The constricting band would slowly starve the blood supply to the little calf's testicles and painlessly usher in a life of steerdom marked by reduced libido and increased weight grain.

"Sure beats the knife," Del said. "But if they're more'n a couple months old, we'd have to castrate. Dog eats well though."

"Me too," Harlan chimed in. "I like a mess of them oysters occasionally."

"Hey! Get away from here!" Del yelled, waving his arm at Maggie, Zach's half mutt, half spaniel that started barking at the immobilized calf. The calf's moist dark eyes widened with fright. "Harlan, see if you can tie up that dang dog before she barks again. This poor little feller's scared enough being here already. Don't need that crazy dog around."

"This calf's got some scouring too," Harlan added, noting a telltale yellow ochre smear around its tail. "Better give him a pill. Then I'll get rid of the dog."

Del tore open a packet of foil-covered antidiarrheal pills, placed one in a small cup-shaped holder at the end of a long plunger, and forced the big pill deep into the calf's throat. With head down and tongue protruding, the calf looked like it was going to spit out the pill. Del reached down and gently stroked its throat a few times. The calf swallowed hard and straightened up.

In less than two minutes, the calf had received a numbered ear tag, two vaccinations, a time-release hormone implant, an antidiarrheal, and a rubber castration band. It was a spring rite of initiation—a coming of age. Despite the brief trauma, Del suspected memories soon faded.

Del usually branded his own calves, too, an additional step in this coming-of-age party, but they weren't branding this morning because these cattle were owned by one of his landlords, the only herd he looked after that wasn't his own. Earlier in the week, Del had accidentally electrocuted a calf while branding because of a short in his electric branding iron. For a moment, when the calf fell to the ground, he didn't realize what had happened. He thought maybe the calf was just frightened. Then, to his horror, it went limp. He called the vet.

"You'd be surprised how often this happens," the vet had told him. "It's your branding iron. Get it fixed or, better yet, get a new one!" It was sobering news, but he knew that whenever you cared for large numbers of animals there were going to be accidents. He just tried to minimize them.

Working steadily through the morning, Del and Harlan processed calves in assembly-line fashion. Once farmers or ranchers might have dehorned their calves as well, a sometimes bloody procedure with the potential for infection or bleeding, more traumatic than anything Del and

Harlan had done today. Now, polled or hornless breeds of cattle such as Angus and polled Hereford had largely eliminated that task. By noon Harlan and Del had processed thirty-eight calves, most of them born within the last two months. The remainder, about forty, would be worked in the afternoon. It had gone slower than usual this morning because Zach and Caitlin were not around to help fill injection needles, number ear tags, or keep an assembly line of tags and pills ready. Yesterday Zach went with Amber to visit her father, now ill with cancer, and they'd stayed the night. While they were away, Caitlin had volunteered to look after Nathaniel. It was his first night away from home. After her long night she told Del at breakfast that being a grandmother had just taken on new meaning.

Often the hardest part of vaccinating and castrating calves was the rounding up and sorting. Sometimes it went well, other times the cattle were difficult to control. This morning it had gone smoothly, in part due to the mineral supplement that Del used to entice the cattle into the holding corral. A few cows, however, hadn't gotten the message, and Harlan had sped back and forth on a four-wheeler, yelling wildly, wheels spinning and throwing up showers of dirt as he chased laggards and errant calves that could run faster than their mothers.

It was a perfect spring morning, brisk and chilly early, clearing and warm by mid-morning. Greening pastures, like immense, manicured golf courses, were moist but not muddy. Flocks of chatty blackbirds, mostly red-wings, streamed overhead in long undulating lines from time to time. Fencerows and woodlands were sheltering many newly arriving birds— wrens, thrashers, kinglets, a few warblers, a dozen kinds of sparrows—all riding the gathering crest of spring migration, a rite-of-spring of another kind, and one that passed largely unnoticed by many humans. It was a trickle this morning, but in two or three weeks there would be a flood of birds pushing northward for a brief rendezvous with the cornucopia of summer. And there were other signs that spring's transformation was underway. Still risking a freeze, brave flowers such as rose verbenas and bird's-foot violets dotted prairies and pastures. They were vanguards. In a few weeks, prairies and roadsides would be sprinkled with the color of dozens of kinds of flowers.

Harlan may have been occupied with cows this morning but the fine weather hadn't been lost on him either. A spring morning like this, in his mind, would have been perfect for fishing. Peak crappie spawning was about two weeks away. Crappie were chunky, humpbacked fish related to sunfish. Their silvery bodies were densely spotted with black, and even though they seldom got much above two pounds, they were excellent-tasting freshwater fish. Harlan considered them his specialty, and he loved fishing for crappie more than anything else. He knew he'd have to get away from these damn cows for a few days, and soon.

Once the cattle and calves were rounded up and confined to corrals, the cows then had to be separated from their calves. It took patience and quick reflexes to separate the nimble calves from their mothers, but once the calves were vaccinated and castrated, they were quickly reunited with their mothers. This was normally the last roundup until fall when the calves were weaned by permanently separating then from their mothers and then sold.

Working calves was a chore, but Del enjoyed it. During a short break, he paused to look over the calves still penned in the corral. They were cute, all about the same size, chunky in shape and with soft, fuzzy black coats that shone in the sunlight. They had sturdy legs and broad heads, and every one had floppy ears and wind-up tails. Among the black ones were a few with white faces betraying a mixed ancestry. There also was a sprinkling of Charolais, which varied from pale gray to almost pure white. Charolais calves were larger and heavier-boned for their age than the Brangus and Black Angus but not as compact and squarish in shape. By the end of summer, the Charolais would continue to hold their edge in weight if their mothers had enough milk, but the Black Angus would command premium prices. Del figured, however, that the higher price was just marketing, a desire fabricated in the minds of advertising agencies and a shrewd Black Angus cattle breeders' association. Restaurants featured "Black Angus" steaks as if they were superior to Hereford or Charolais or other breeds. Somehow "Charolais steaks" didn't have the same cachet. Del wondered if people just didn't know how to pronounce the name.

According to Del's records, thirteen of the ninety-one cows in this herd had not yet had calves. These cows were separated from the main herd and put in a small nursery pasture. Separating them now would eliminate a second roundup later, and when they did have their calves they could be quickly processed and moved back to the main herd.

There were more cowherds to work, but the present batch was one of the largest. With Caitlin and Zach's help, they could finish the other herds in four days, maybe less.

By Saturday afternoon, the sixteenth, Del started planting corn. He worried about corn because it was famously drought sensitive, but the newer, shorter-maturity corn hybrids, he figured, had a better chance of maturing before a late summer drought if planted early. The risks of planting too early were cool soil temperatures that might delay or prolong germination and seedling development, but it was a risk he'd take. For the next week or ten days, he would think about little but planting corn. His worry was a rain delay. It was always the final arbitrator of spring planting.

15

May 8
Storm

May. Farmers finish planting corn if weather delays April planting; soybeans and grain sorghum follow; fescue mowed and baled for hay late in month; farmers spray and cut thistles; herbicides applied to emerging crops now through June; lightning bugs appear on warm evenings; strawberries ripen in gardens; spring wildflowers bloom on prairies and roadsides; birdsong peaks by end of month.

Del's truck rolled to a stop just short of the farm repair shop. He jumped out and walked through the small wooden side door. Zach and Harlan were inside. "Either too wet or too dry," he said to neither of them in particular. "Looks like we got more rain coming today. At least we got all our corn planted a couple weeks ago. Been a mess working around the rain."

Zach was mounting tires on a two-wheel trailer. Ignoring Del's weather predictions and without looking up, he said, "Hey, Dad, I can't get the electric fuel pump to work on this bulk fuel tank. Would you take a look at it?"

"What's wrong?"

"Works when plugged into 110-volt AC but not when I use the DC-AC converter connected to a battery. Just hums, then stops."

Del experimented and got the same results. "I'll call tech support. See what they say," he said. "Probably get somebody in China answering the phone."

"Don't matter, I guess, if they can solve the problem."

With tractors and combines burning five to six gallons of diesel an hour, Del burned more than two hundred gallons some days. He had been refueling in the fields from an auxiliary tank on one of his pickups, but the tank was small and the pump slow. That was when Zach decided to build a trailer from spare parts and mount a three-hundred-gallon tank on it. He even got some free green paint for it from a friend when a boat manufacturing company moved.

While Zach finished mounting the tires, Harlan ground off some green paint on the trailer's hitch with a portable grinder. The grinder sent a cascade of orange sparks bouncing across the shop floor toward Boy the shop cat, who leaped clear of them and then stared at Harlan as if he had just suffered a great indignation. Harlan then welded a jack to the bare spot on the hitch. The jack and fuel pump were the only parts they'd had to buy.

Del opened the large sliding shop doors to let some fresh air in and clear the haze of welding smoke and grinding dust. As he slid the doors open, a rain of dried grass and feathers from a house sparrow's nest hit the concrete. Some of the debris fell on his shoulders and cap. Brushing it off, he shrugged, "You know, I open this dang door almost every single day, and their nest falls down, and they still build it back. You'd think they'd learn."

"Or maybe you ought to learn not to stand under it," Zach said, laughing.

Outside the shop, Del loaded an old 300 series Massey Ferguson combine onto a flatbed trailer. He'd bought the combine at a farm sale because he wanted the pick-up attachment mounted on the header. It was an Allis-Chalmers add-on with new belts and pick-up springs. He also kept the battery. The rest of the combine was junk, so he called Crockett's Machinery Sales in Grand Forks, and the owner, Carroll Crockett, offered him three hundred dollars for it sight unseen—scrap iron price and almost what he paid for it—if he'd bring it in.

Outside, Harlan helped Del load the old combine. Pausing to look at the buildup of oil and grease and dirt on the machine, he exclaimed, "Man, it's a wonder this thing ain't burned up already."

"Some did," Del said. "Get this much grease and oil around all these moving parts and sooner or later you're gonna get a fire. More'n a few guys burnt up their wheat fields that way." He climbed up the combine's ladder and wired the broken cab door shut, then attached red flags to the sides of the header, which extended beyond the width of the trailer. "Look at this engine. Massey put a little Chrysler Slant-6 engine in it. Wonder if it still runs?"

The question was rhetorical, but to Del the common denominator of everything mechanical was parts. He'd let the other guy buy new. He almost always bought second-hand and fixed his equipment with parts cannibalized from older machinery.

Del glanced up at the sky, an automatic gesture. He could see a storm was brewing. A blackish cloud, sculpted smooth and round on top and flat on the bottom, loomed low in the southwest. He wanted to get the combine up to Crockett's sale lot for their upcoming auction, and he could take Zach's refurbished moldboard plough along too. The sky didn't look right, and he knew it, but ignoring his better judgment he went to the house to get his wallet. That was when Caitlin found out his intentions.

"You see that sky?" she scolded. "We got a storm coming. Don't you go and leave me here alone with Nathaniel until things clear up." Caitlin was sanguine about most things on the farm. Too many seasons had come and gone for her to get worked up about little problems, but she was rattled by storms recently. Tornados in Missouri and Kansas had leveled two towns and destroyed parts of several others just two days ago. Nearly thirty people had died in the Kansas tornado. It was the worst in years.

Harlan echoed that fear too. "You hear about the town of Stockton?" he'd asked her yesterday as he jumped out of his truck and burst excitedly onto the back porch. "People's saying it's gone! Just gone! They ain't anything left! Can you believe it?"

"I heard that," she told him. "It's all over the news." Caitlin guessed Harlan was so upset because he often went fishing at a lake just outside the town. He would have known lots of people there.

Then, turning to Del, she said in a firm tone, "If it gets worse, I'm going to the storm cellar with Nathaniel. You and Zach ought to come too."

Del hesitated, fishing for excuses. "It's dark down there. Don't smell good neither. And besides, there's water on the floor."

The old cellar was leaky. Built to store vegetables and perishable food before electricity reached rural areas, it was a small rock and concrete structure sunk about four feet into the ground and covered with a rounded mound of earth for insulation. A lift-up door gave access to steps leading down to a second door and the storage area. Del and Caitlin had never stored anything in it. He patched a leak in the cellar's wall once and it was dry for a while. Now it was leaking again and the drain was clogged. The entrance door leaked too.

"I know there's a little water on the floor, but it's not that dark. Light comes in from the air holes," Caitlin countered, hoping to sway his decision.

Del knew Caitlin had a point. He shouldn't leave her and his grandchild when a storm was gathering, so he said he'd stay. He joined Zach and

Harlan outside; they had finished with the fuel trailer and were now disassembling an old school bus temporarily parked next to the steel grain bins. With all the rain and not being able to get into their fields, Del and Zach had gone to some sales instead. Caitlin had been surprised when they came home with the bus. Del figured to gut the inside of it and put in wooden shelves to store machinery parts.

Zach said there were some mechanical parts from the bus that he could use and was explaining to Harlan what to save. "Take out the wiring harness, gauges, steering column, everything under the dash," he said.

Harlan, on his back under the dash, face contorted, grunted as he sorted through the wiring.

"Maybe pull the engine, too, if there's time," Del added. "Probably do as a spare for one of our trucks."

The sky was dark and even more ominous now. The long, low cloud in the south had the appearance of some kind of supercell, Del thought. He'd seen pictures of those on weather reports. Not a blade of grass trembled. Nor did a sound issue from the martins huddled on the rim of their birdhouse apartment, nor from the dickcissels in the pasture. A distant bobwhite whistling from atop an old fence post went silent and hopped to the ground, slipping away quietly. Fescue grass billowing in the morning breeze moments earlier now seemed frozen in time, the new seed heads, scarcely in boot, were stock-still. The air was moist, charged, as if the impending storm was silently building up its own internal energy.

Storm sirens wailed in Prairie Point three miles away, and in Bellamy twelve miles away. Their distant moaning echoed across the countryside. Caitlin heard the sirens, grabbed Nathaniel, and hurried outside to find Del and ask him, for the second time, about going to the storm cellar. He said he thought it would be better if she just waited outside with them so they could all watch the weather. Satisfied, she brought out an aluminum folding chair and sat along the west side of the machine shop where she had a good view of the evolving storm and the men working on the bus just across the driveway. Holding Nathaniel on her lap, Caitlin was uneasy about being outside, but there was comfort in being close to the men. It suited Del too. He didn't like the idea of getting in the storm cellar, but agreed to go if he saw a tornado approaching. Del guessed that from their outdoor vantage point they could see in all directions. It began spitting rain and the wind kicked up.

Zach's cellphone rang. It was Amber.

"Hi," he said, trying to sound upbeat. "You about to blow away down there?"

Amber, working in a doctor's office in Bellamy, had heard the sirens, too, and was worried about Nathaniel. "What's Grandmother Caitlin doing with Nathaniel? Are they in the storm cellar?"

Zach hesitated, downplaying the danger. He told her they could see the storm but it was passing to the south of them. "And yes, they'd get in the storm cellar if the storm moved in their direction." He didn't tell her that at that very moment he could see a massive lightning storm visible in the interiors of several huge, globe-shaped thunderheads. For an instant they were illuminated like immense Japanese lanterns. Nor did he tell her that he could see a huge slanting wall of rain, gray blue and opaque in the distance, that looked like it was headed in her direction. He hesitated, knowing she was worried, and he didn't want to increase her anxiety. Still, he needed to warn her just the same, so finally he said, as calmly as he could, "You be careful too. Looks like a big rainstorm is headed in your direction."

Fifteen minutes later, the sky began to lighten and the storm moved off southeastward. Soon dickcissels were singing again and martins were burbling and chortling and swooping back and forth in loopy circles across the driveway. However, another line of clouds lay on the western horizon, still far away, but low and dark like an ominous frown.

With a break in the weather, Del and Zach took the combine and plough to the sale lot in Grand Forks. Zach had paid fifty dollars for the plough at a sale, welded on some broken parts, and painted it green using his free paint. Small ploughs were in demand for gardens in urban areas, and unlike big four- and five- and six-bottom ones farmers once used, these little ones would get bid up at sales. Most farmers had now shifted to minimum-till or no-till methods, a quantum change from Del's childhood. Zach had ploughed Caitlin's garden with the little plough last week. Earlier in the year, he'd built a three-bottom plough from spare parts and scrap iron. It brought two hundred dollars. This one, a factory-built Deere, ought to bring more.

A company auctioneer tagged the plough and motioned for an operator to hoist it into one of the long lines of farm machinery being organized for the monthly auction. Inside the sale building, a machinery buyer named Sid peered out the window at the old Massey combine loaded on Del's flatbed.

He looked at the rust, faded paint, grease-soaked panels, and cracked tires, eyeing it all for an uncomfortably long time without saying anything.

"Carroll told us he'd give three hundred for it if we'd haul it up here," Del eventually interjected, hoping to get the conversation moving. Carroll Crockett was the owner and patriarch of the auction company, and Del figured mentioning his name might add credence to the offer. Carroll had made the offer verbally, and he wondered if Sid would stick to it. Where the heck was Carroll right now? he wondered.

"He did, did he?" Sid mumbled, his face stony as granite.

Del suspected the face was a ruse.

"Wonder what he was thinking about?" Finally, after another pause, he drawled, "You boys want us to just credit the three hundred to your account?"

Del nodded, the deal done.

When Del and Zach returned home, they discovered that Caitlin also had decided the weather was safe for a while and she'd left for a hair appointment, taking baby Nathaniel along and leaving warm food on the stove for the men. The aroma of roast beef and mashed potatoes infused the kitchen. After their meal, Del, Zach, and Harlan lingered to watch weather reports coming out of Springfield. Another storm front was on the way, likely the one they had seen far off to the west earlier in the day. Del listened to Harlan fret about an upcoming crappie tournament. He and a friend had put up the four-hundred-dollar entry fee, a considerable amount, but the prize money was good. He hoped the storms didn't postpone it. He explained that with the water warming, the female fish would soon be spawning. Once they spawned, any females they caught would be lighter in weight—and that wasn't good in a tournament where winning was based on the cumulative weight of your top ten fish. Del pointed out that the other fishermen would be similarly affected, so it would all balance out, but Harlan didn't see it that way. If the tournament was postponed, spawning would surely be finished, and he didn't want that to happen.

Harlan and his friend had been laying crappie beds over the winter, mostly old branches they submerged with rocks to keep them hidden below the waterline in the lake. They took GPS readings of the locations so they could find them again during the tournament. Before they got a GPS, they'd used little red floats to mark their hidden beds, but in an earlier tournament a fisherman found some of their floats and fished their spots first. Harlan was damned determined that wouldn't happen again.

After their midday meal, Del told Zach about a cow with foot rot. "You better treat her before the storm comes."

Zach picked up the crossbow they sometimes used to administer vaccinations to cows. With it he could shoot a self-injecting syringe loaded with penicillin and treat the cow without rounding her up in a corral chute. The cow was with a herd a few miles north and west of the homeplace. He glanced at the sky as he got in his brown pickup truck. Even with a storm coming, he should have time.

For a few minutes the sun broke between heavy clouds and brightened soggy fields. In wet pastures, light reflected off water droplets like millions of tiny mirrors in the early afternoon sun. But the air was hot and oppressive, and Zach could see the sunshine wouldn't last. It was merely an interlude between warring storm fronts. Clouds piled up in the west, great vaults, domed, blackish, inching closer. In the pasture Zach drove through a small creek, angled up across an open slope, and, at a gated corner, entered another pasture. The cattle were at the back of the second pasture, some of them standing beneath a triad of Osage orange trees. Bunched under trees in open pastures, the cattle were little more than electrical conductors underneath living lightning rods. But cows did that. So far this year they'd lost three cows to lightning, one along a fence, two beneath a tree. He knew a neighbor that used to have dairy cows. Years ago lightning hit his barn, but it didn't follow the lightning rods. Instead it traveled down the tin roof and killed seven cows at the corner of the barn, splintered wooden beams, even tore off part of the roof. The neighbor found his cows the next morning, some of them on their backs, blackened legs out stiff and straight. Sometimes a year or two passed without a killer strike. Last year another neighbor lost fifteen beef cattle in a single strike. He didn't have insurance. This year he did.

To an onlooker, it would have looked strange—a man leaning out the window of a moving pickup truck and aiming a crossbow at a cow. The crossbow's arrow was a fat cylinder with a hypodermic needle and syringe in front and a compressed-air cylinder behind. Zach paused to hand pump several pounds of air into the cylinder. When the needle penetrated a cow's thick hide, the compressed air cylinder instantly injected the medicine in the syringe. Then a spring-loaded plunger ejected the needle and cylinder, which fell harmlessly to the ground. Ringed groves on the needle kept it from slipping out for a second or two. It worked well the first time or two on a cow. After that most cows wised up.

Zach spotted the cow. She'd been treated previously with the crossbow and would probably remember the unpleasant surprise. His suspicion proved correct. He couldn't get close enough to fire the syringe-tipped arrow out the truck window. He tried on foot, then circling in the truck, then again on foot, crouching and hiding behind other cattle. The cow was wily and kept her distance. The herd milled, heads up, ears forward, a line of noses like smooth wet softballs. They eyed the man in their midst with suspicion. The cow, limping, kept a buffer of cows between herself and Zach, and that prevented a clean shot. If she was in the open she was too far away. The cat and mouse game continued. Suddenly, a gap appeared, the bowstring hummed, and the syringe found its mark and then self-ejected. Zach walked over to retrieve the spent syringe and arrow, rubbing a little dirt off the ringed groves on the needle. The cattle were more dispersed now, uneasy with the man behaving oddly. This cow had been cagey, but it had been easier than a roundup.

By the time Zach returned to the homeplace, the skies looked threatening again. As before, it started with an eerie calm, as if the land was holding its breath. Del decided to stay outside and do cleanup work where he could watch the sky. Behind the shop an archipelago of rusting scrap metal, most of it jetsam from winter projects, lay in piles in knee-high grass. Some of it projected like jagged metallic islands above the sea of grass; the rest lay submerged, a peril underfoot. From time to time, Del hauled his scrap metal to a corner of his eighty acres east of town where it would be out of sight. He told himself if prices rebounded, as they did last winter, he'd sell it, but he hadn't found time to get rid of it. The flatbed trailer was empty now as Del and Zach began loading the scrap metal. Harlan wasn't around. He'd been sent to work at one of the rental farms.

The two men loaded the heaviest pieces of steel with a hydraulic loader on a tractor, in some cases by wrapping a chain around the pieces and attaching the chain to the loader. Lighter items were tossed by hand into the wide bucket loader on the front of the tractor, then dumped onto the trailer. It was heavy work, lifting, dragging the steel and iron, tugging at large pieces in the grass, wrenching them free with noisy clanging and banging using the hydraulic loader, all the while keeping a wary eye skyward. They watched as the calm gave way to a breeze, scarcely noticeable at first, then stronger, roiling knee-high grass that swirled and flashed silver and green semaphores across fields behind the farm buildings. Swallows beat hasty retreats to sheltering barns. Four crows tumbled across a gap in the wooded stream south of the pasture, half flying, half pushed by gusting winds as they, too, sought shelter. The mockingbirds and the wrens and the martins fell silent again.

Del paused to answer his cellphone. It was tech support returning his call about the DC-AC converter that was frustrating Zach. They asked about the length of the electric cord on it. "Try using a shorter cord," the voice crackled, the wind making it difficult to hear. "No more than twelve feet in length; six feet would be even better."

"It was an eastern voice," Del said later. "I could tell by the way he talked." He was suspicious of their advice but he'd try it.

Del and Zach continued working. Sirens wailed for the second time that day. Caitlin, back from her hair appointment, looked at televised weather reports with apprehension. The sudden strength of the storm surprised her. Forecasters in Springfield now were showing the approximate center of the storm front maybe ten to fifteen miles to the west of their location. She realized they were directly in the storm's path with only minutes to maybe a half hour to prepare. Fear crept into her stomach because now she could see the storm approaching—the forecast was like a glimpse of the future. She watched the blackening sky from the kitchen window, which faced to the west. It was darker than a few minutes ago. An immense, gray-black cloud at the leading edge of the front curved away in a gigantic arc to the north and south. A neighbor four miles to the west called to say it was hailing and winds were gusting strong. She told Caitlin to take cover but said nobody had seen a tornado. That was enough for Caitlin. With Nathaniel wrapped in a blanket, she hurried outside to find Del. "We need to get in the storm cellar!" she called out as Del was putting a tractor in a shed.

While Caitlin was talking, Zach's phone rang. Amber was following storm developments from work. "Please, Zach, take Nathaniel and get in the storm cellar," she begged. He said he would if the winds got bad or they spotted a tornado.

"The Springfield weather service is showing the storm center west of us," Caitlin said, her voice urgent, pleading. "We need to get in the storm cellar."

"What are they saying?" Del asked, not understanding her in the wind but showing concern.

"No tornados have been sighted, but they're advising everybody to take shelter," Caitlin replied more forcefully.

Del decided to accompany her back to the house, but he bypassed the cellar. "We can watch from the house for a while," he said.

The air was suddenly heavy and warm, and everything was bathed in dark yellow light. Zach stayed outside, spellbound by the dramatic transformation of the sky, as if he were witness to an approaching apocalypse.

Del glanced at him. "If you and Caitlin don't go down in that storm cellar, then you two better get your stories straight so you can tell Amber the same story when she gets back tonight," he said with a laugh. "She's going to want to know what you did."

Zach said he would, then lingered outside a little longer. Caitlin followed Del inside the house and they watched from opposite windows, the sky ominously dark. Outside, Zach watched the low black clouds churn and move with alarming speed. In the distance some were narrowly illuminated in silver, as if they had giant halos. Moments later, a black, funnel-like shape expanded downward. It was smooth and sculptured. His pulse quickened, and he thought he was seeing the birth of a tornado. The vortex lingered for a few frightening minutes, then seemed to abort, merging into the dark mother cloud above.

Crissy paced back and forth, dragging her silver chain attached to a stake. The wind upended her doghouse, sending it cartwheeling across the garden and pinning it against the fence at the far end of the garden. Zach ran to grab it before another gust could send it spinning across an open field. The top came unsnapped and both pieces were jammed against the fence by the force of the wind. Grabbing both sections, he struggled, walking backward against the wind, the two pieces flailing in his hands, until he reached the shelter of the shed and pushed the pieces inside, slamming the door.

Distracted by the storm and wanting to warn his parents if a tornado approached, Zach forgot about the dog still chained by the shop. He hurried out to the corral where there was an open view of the skies. Leaning into the wind, he watched as a herd of cattle, with heads down and tails to the wind, stood in a resolute cluster by a gully in a pasture. Suddenly hail began hitting the ground, at first just a few pieces, then a drum of near walnut-sized chunks, as if disgorged in a spasm. A few of them hit his head and back. Some were round, others irregular and broken like fractured shards. Hurled by the wind, they stung his face and arms, and he turned and ran toward the back porch, hail and rain pelting his back, wind gusts pulling at his clothes. As he ran, he saw Harlan turning into the driveway, gunning the pickup truck toward the shop, then skidding to a stop in the gravel. Harlan had been working on one of the landlord's farms and overstayed. Eyes wide and ashen face a panic of lines, he leaped from the truck and ran, hands over his head, for the machine shop. He'd hoped to back his boat into the shed but had waited too long. Now his boat was turning into an oversized ice chest. A riot of hail shattered like glass on the driveway. He reckoned his fishing trip this evening was finished.

Zach ducked into the house. Wind slammed the house with blasts so strong it shook the walls. Caitlin wanted to go to the storm cellar.

Del grinned calmly. "Too late now, I guess. You want to go out in that?" he said as he motioned with a wave of his hand toward the window. "Might as well stay here unless we actually see a tornado."

Looking out a back window, Del saw Crissy. The dog was huddled against the side of the machine shop, the wind whipping her long black and white coat into whorls. Blinking from the rain but sheltered from the worst of the storm, she sat with her back to the shed as the fury of the storm swirled past. Del wished he'd thought to put her inside the shed although he knew she'd be fine where she was, just wet.

In the house the three of them watched the storm from different directions, anxiety showing more on Caitlin's face than on the men's. A hard wind buffeted the house, rain rattled on the roof, and they watched and waited, and wondered about Harlan, riding out the storm in the shop. The shop's tin roof would make it sound worse. The storm was violent but relatively brief, perhaps no more than twenty minutes, but it seemed much longer. Soon the rain slacked and, with the wind squandered on the storm, it fell calm and cool.

After the storm passed, Del ventured outside and unfastened the dog. She was wet but overjoyed to see him, repeatedly leaping up with uncontrolled enthusiasm, muddy paws raking Del's pants. Then she shook herself vigorously and bounded away across the soggy yard.

"At least we got a clean dog now," Del said, laughing as he saw Caitlin appraising damage to the garden, "but that won't last long." He checked a plastic rain gauge nailed to a fence post beyond the shop. Almost three inches. The air was fragrant, charged from the storm. With the rains last week, it would be another week before they could get back in the fields to plant—and that was if it didn't rain again.

16

May 18
Thistles and Thieves

Days were lengthening, and there was no longer a chill in the morning air. Frost-tinged mornings had given way to dew on ankle-high grass. Across rural Mid-America, tractors were being fueled and readied. Machinery was being brought out from winter hibernation in dusty old barns and greased and inspected and given a clean bill of health for another season. Hydraulics were tested, too, heavy cast cylinders with shiny, mirror-smooth shafts that did agriculture's heavy lifting—the raising and lowering of fleets of farm implements. That so much force could be generated by an oily fluid inside a few inches of flexible hose had much to do with the application of a physical law discovered by a French mathematician nearly four hundred years ago. Pascal's law, largely unheralded, had played a key role in the modernization of agriculture, and the evidence was everywhere. Hydraulic hoses joining tractors and machines like black umbilical cords are now indispensable lifelines powering everything in agriculture today. The pungent odor of sweat and leather and horse manure of the past has today been traded for the fumes of diesel and red hydraulic fluid.

While horses and manually operated machinery ignited an early revolution in agriculture, it was one built largely on brute strength and cheap physical labor. Rural populations were large then, and individual farms small and inefficient. Today rural populations are small and farms much larger, in no small part due to the use of hydraulics, which were first adopted in the 1950s. Likewise, the flies and fleas and unwanted insects and weeds of that same vintage rural America are now largely traded for an arsenal of chemicals in agriculture. And they work even while farmers sleep, holding in checkmate an army of insects and weeds we have designated as pests. This morning Harlan was about to put a sample of that chemical arsenal to work.

For the past two weeks, sporadic rains had kept fields too wet to work. Still, farmers found work to do—looking after cattle, grinding cattle feed, cutting and spraying brush, fixing fences—but not what they wanted to do. Del watched prime planting days slip away. Caitlin reminded him that the year before, also a wet spring, it was the first week of June before he started planting milo and beans. And she pointed out that the crops still did well, even if they were late getting planted.

"But last summer we had plenty of rain," Del countered. "Can't always count on that. Dry summers and late-planted crops are a receipt for disaster." He knew they'd been lucky. A failed crop season would be a huge loss—fertilizer, fuel, herbicide, seed, and labor on 900 acres of still-unplanted land. Crop insurance would insulate them from disaster, but it wouldn't replace a good harvest. They'd need almost two weeks to finish once fields dried. If planting extended beyond the first week of June, it would interfere with fescue swathing, and fescue harvest.

Del discussed with Caitlin the possibility of buying a futures contract on their bean crop, or at least on part of it, as a hedge against falling prices. He called a broker at the Kansas City Board of Trade, but, in the end, they decided to wait and buy only if the futures price went above eight dollars a bushel. So far it hadn't, a worrisome hint that prices would be lower this year. Del hated to hedge a crop he hadn't planted. Selling something before it was planted—assuming he'd get it planted—went again his nature, even if it probably was the smart thing to do. The price was good now, as high as it had been in twenty years, but the current market price was higher than the futures price, a bad sign. Caitlin said she'd watch for news that might affect price direction. The broker would keep his eyes on technicals and overseas news. Del and Caitlin knew how globalized farm commodity prices had become. It was different when they had started farming over thirty-five years ago. Farmers today were like flotsam on a sea of international commerce, with little real leverage and no economic safety net except for government price supports. Now it was increasingly difficult to compete with low-cost foreign producers. Farmers in the United States had, for decades, enjoyed dominance in export crops like soybeans, but that was changing. Some of the problems were political, mistakes that hung the American farmer out to dry. Others had more to do with third-world countries catching up, bringing more land into cultivation, developing markets, and challenging the U.S.

for market share. Brazil and Argentina had brought hundreds of thousands of new acres of soybeans into production in the last two decades. Their land and labor costs were far below that in the U.S., but they lagged in efficient transport to shipping points. Sometimes, also, it was speculation by traders that caused prices to soar then plunge, and unpredictable prices were a farmer's nightmare.

Del was stoic about weather delays over which he had no control. He used downtimes to work on machinery, anything that would reduce the chance of a delay once fields were ready. Yesterday he tested a field across the road from his house for dryness. He disked a little semicircle at the edge of the field, and the lugs on the tractor's duals cut deep, smooth furrows in the soft earth and hurled up heavy chunks of wet soil like missiles behind the wheels. He raised the disk gangs, hydraulic cylinders squealing in protest, and then the transport wheels on the disk cut deep ruts. Planting would have to wait.

Harlan was on thistle- and brush-spraying patrol. With a twenty-five-gallon spray tank filled with brush killer and a long-handled spray boom strapped to the back of his four-wheeler, he zoomed along fencerows and made erratic forays out into pastures, swerving, bouncing, and skidding to stops—a hunter of brush and weeds with spray boom and chemicals. Bull thistles and multiflora rose and tree sprouts in fencerows were among the targets of his spray nozzle. Keeping fencerows free of brush was an unending task, but it hadn't always been that way.

Before Europeans settled North America, fires swept tallgrass prairies and kept in check pioneering trees and shrubs such as eastern red cedar, smooth sumac, blackberry, black cherry, wild plum, pin oak, bur oak, hackberry, slippery elm, and roughleaf dogwood. These woody trees and shrubs were adept at colonizing the prairie grasslands despite the perennial heat, cold, drought, and wind. But they could be subdued with fire. It was an epic struggle, a dendritic wrestling match pitting prairie grasses against woody shrubs and trees, and for thousands of years it was largely a stalemate. The battle between sturdy prairie grasses and pioneering eastern trees and shrubs along a midwestern battlefront stretching from southern Canada to Texas dated back to Pleistocene ice ages and perhaps earlier. When Native Americans moved onto North America prairies, perhaps 8,000 or more years ago, prairie grasses may have gained an ally. These plains people learned to use fire for hunting and to gain advantage over adversaries. It didn't end the war between prairies and forests, but it may have shifted the contested zone eastward, perhaps close to where Harlan worked this morning. Harlan's efforts were, in a way, an extension of that struggle—one now distorted by modern human intervention.

Europeans settled American prairies in the 1800s, arriving by the tens of thousands. Determined and eager to work, they set about subduing the prairie biome and building homes and futures. Prairie fires, some so vast and fast moving they were reported to travel forty miles in a few hours, were a terror to these new residents. Then, over time, open prairies gave way to cultivated crops, and fires ceased to threaten. The new settlers replaced diverse native grasses—the blue stems, switch grass, Indian grass, gamagrasses, and sloughgrass—with monocultures of non-native forbs and grains, and they replaced bison, a native ungulate, with pliable domesticated livestock.

Settlers soon transformed a sea of prairie grasses into a checkerboard of domesticated fields and pastures. Roads bisected the prairie, homesteads dotted the rolling land, and a tallgrass prairie biome ten thousand to hundreds of thousands of years in the making became, in little more than a half century, Mid-America's farm belt. This newly arranged landscape was soon organized by fences, a symbol of established agrarian order. Fences confined livestock, defined land ownership, separated neighbors, and made good neighbors better. Fences also provided safe zones of conduct for the germination of errant seedling trees and shrubs, a web of corridors from which resilient, pioneering sumacs and plums and elms, all natives, could launch their unceasing invasion of prairie soils.

For a while prairie grasses found allies in horses and mules, which early settlers brought with them. When these animals weren't laboring in fields or pulling buggies or being saddled for weekend pleasure, they spent idle time browsing sprouts. Unlike cattle, which didn't browse on woody vegetation, pastures and fencerows with horses and mules around were seldom brushy. Hay cutting also kept prairies free of invading brush. Then the balance shifted. The horsepower of horses was replaced by the horsepower of tractors, gradually at first because early tractors were slow and primitive but, by the end of World War II, draft horses and mules were on their way to obsolescence for farm work. Horses became entertainment, mules a novelty, and within a few generations the once treeless prairie was transformed. Today it was Harlan and his spray nozzle that stood between a few prairie fencerows and pastures and a hostile takeover by a fecund array of native brush and alien exotics.

Settlers from Europe and Asia also brought with them new plants and seeds, a cornucopia of germ plasm set loose on a virgin landscape. Some of the introductions were unintentional, among them a botanical rogues gallery of stowaway seeds and spores transported across the ocean in the guts of their livestock, in the hay and grain fed to livestock and secreted in their food, clothes, shoes, body hair, farm equipment, and personal items. Others were brought purposefully—flowers, medicinal plants, and edibles that had been important in their lives in their homelands. Even the cursed bull thistle may have arrived with someone's blessing because of its reputed association with medicine that reached back two millennia. Its generic name, *Cirsium*, meaning "swollen vein," hinted at the role of thistles in ancient medicine. A related *Carduus* thistle once played a cameo role in Shakespeare's *Much Ado About Nothing* when he wrote, "Get you some of this distilled Carduus Benedictus and lay it to your heart; it is the only thing for a qualm." Legend has it that in Scotland thistles piled on beaches helped repel a Viking invasion. For its service, the thistle, forever after, occupied a place of honor on Scotland's heraldic emblem with the motto "Touch me who dares!" It is a lesson worth recalling. Thistles are so profusely armed with spines they repel grazing livestock.

On this side of the ocean, the bull thistle enjoys no such exalted status. Instead, it has landed on virtually every state and federal list of noxious weeds as one of the most insidious invasives. Disseminating millions of tiny wind-born seeds, bull thistles crowd native plants and degrade pastureland, rendering them useless. Of nearly fifty kinds of thistles in North America, almost half are aliens, but few have incurred the wrath of landowners as has the prickly bull thistle.

Harlan had other weedy aliens in his sights too. Multiflora rose, originally from Japan, was imported by rose growers who grafted its roots to cultivated varieties of roses, but it had an unforeseen dark side. Some botanists described it, euphemistically, as "vigorously colonial." Others said it was a pest. *Rosa multiflora*, like all true native wild roses, has only five petals, not the many overlapping rows of petals found on cultivated roses, and its white blossoms are attractive and resemble those of wild plums and blackberries. Its red hips are eagerly devoured by mockingbirds. In the 1950s, when Del was young, it was heavily promoted by conservationists and agricultural planners as good wildlife cover and as a living fence. Landowners planted thousands of miles of them. His father and many of his neighbors planted it. Confined to fencerows, the rose was good wildlife cover. It also was a good fence. Then it spread, escaping beyond fencerows in a transmogrification so complete that those who once planted it eventually fought it with bulldozers, fire, mowers, brush hogs, herbicides, even salt, as they watched it invade and cover pastures, old fields, even woodland borders.

While Harlan rode his four-wheeled steed on brush and weed control, Del checked cowherds, then took a fourteen-ply tire to town for repair because it was so stiff it required professional equipment.

Zach, meanwhile, loaded two cows and their new calves from the maternity pasture by the homeplace into a cattle trailer and moved them back to their herd. One cow was new to the herd, the other simply absent for a few weeks while she awaited the birth of her calf. Now they were "new kids on the block," humbled before a bovine sorority that knew its members and tolerated no admittance without initiation or reinitiation. The herd rushed them in a seriocomic burlesque of jostling, pushing, snorting, sniffing, and tail swatting. Panicked, the two cows bolted, then tried to return to their disoriented calves, but they were reduced to fending off aggressive shoving by other cows intent on retaining the status quo in the social order, and keeping a discreet distance from an amorous, semi-aroused bull's vaginal sniffing. But attention spans were short, and social hierarchies, such as they were, soon realigned. Within minutes, the focus returned to the more mundane bovine business of grass eating, fly swatting, and cud chewing.

When Zach left the pasture, Owen, the landowner, shuffled out stiffly to greet him and open the gates by his house. The fence next to the gate was garlanded with white blackberry blossoms, and when the men stopped, a startled mockingbird flew up from amidst the blossoms, flexed its wings upward to reveal startlingly large white patches, and cocked its head to stare. Owen had just returned from a brief hospital stay. A retired military pilot with intense blue eyes, he was lean and ramrod straight despite his seventy-five plus years.

"It was such a nice morning I thought I'd walk out to my orchard and spray some apple trees. Laid up like this I can't do much else right now," he said, "except maybe keep an eye on things."

"Anything wrong?" Zach asked, sensing something might be troubling Owen.

"Thieves got in here while I was in the hospital," Owen said, his voice hard, tension showing on his face. "Stole my chainsaws and an air compressor. Siphoned the gas out of my tractor and even the garden tiller. Looks like they tried to siphon it out of the truck, too, but couldn't get the hose in it. The cap was off. I'd like to get my hands on that bunch!"

"Lot of people been having problems," Zach said. "The Henderson place just north of you was robbed too. Sheriff has an idea who did it. One of them already has a prison record, they say."

"I saw where your dad and J.P. put up a cable across that driveway. It won't stop 'em, but they'll have to drive through the ditch now."

"That's the second time they've hit that house," Zach said. "J.P.'s mother's been in the rest home for several years now, so nobody's there. The first time they took the TV and some small stuff. This time they took all the furniture. Cleaned it out."

"J.P. shouldn't have left all that stuff in the house. But these guys are getting bolder," Owen said, his eyes narrowing. "By God, if I ever catch 'em around here again, I may just ventilate their damn vehicle with my shotgun."

"Don't blame you," Zach said. "It's a shame when you can't leave your place for a few days without worrying about somebody stealing stuff."

"When I retired here nearly twenty years ago, we never had problems like this. Lots of people didn't even lock their houses. We moved out here for what rural life is supposed to bring. Good neighbors and a peaceful life. The neighbors here are still good, but there's some types in these little towns that aren't. Sometimes I see people driving real slow up and down the road here. Probably looking for somebody not at home."

Later in the morning, at the farm shop, Zach repaired a metal calf feeder and Del rewelded some brackets that fastened the rear wheels to the axle of an old Farmall M. When he bought that tractor years ago, he'd fixed it up and farmed with it. Later, when Zach got interested in tractor pulling, he bought another M, and the two of them tinkered enough with the engine of that one to double its horsepower. For a while Zach out-pulled many of his competitors at pulling meets and took home armloads of trophies. Then others made adjustments and caught up.

Caitlin appeared in the big shop door to tell the men that dinner was ready. Offhandedly, she announced, "Did you know that Crissy can fetch?"

"Ha! I never seen that dog show any interest in anything but eatin' and ridin' in a truck," Del said, "and maybe chasing cows when you don't want her to."

"No, look," Caitlin said, throwing a small rubber ball. Crissy immediately chased it down, skidding in the dirt as she grabbed the ball, and hustled back to Caitlin.

"Let me try," Zach said, throwing the ball much farther. Crissy sped after it, then stopped several yards short of the ball, seemed to forget her mission, and returned without the ball.

Del laughed. "Just like I thought! That dog can't remember anything for more'n a few seconds. Let's go eat."

Caitlin had deep-fried Harlan's catch of crappie from his tournament a week ago.

"These were the biggest ones we caught," Harlan said, pleased with himself. "We finished in the top ten. Unfortunately, we weren't in the money. Only the top three get money."

"Got any more tournaments coming up?" Del asked, hoping that Harlan wouldn't decide to take off more days in the middle of spring planting.

"There's two more, but I don't think I'll enter. Figure we'll be in the middle of planting and fescue harvest here and you'll need me."

"If the rains hold off, next week will be our busiest of the spring," Del said.

"My summer Bible school class for kids is coming up third week of June," Caitlin said. "Looks like the timing will be bad if you guys are in the middle of fescue harvest. Anyway, it's just the preschoolers for a few hours in the mornings. I'll be around to help out most of the day if the post office doesn't call."

When the conversation turned to weather and field conditions, Caitlin said she'd put out thirty-six tomato plants the day before and her garden seemed to be drying nicely, especially with the changes Zach had made to the garden's drainage.

Zach decided that if the garden was drying he ought to check their fields again while Harlan continued spraying. Del said he needed to pick up some sheet metal and farm supplies in Ashton.

By early-afternoon, Zach called Del and said that a couple of fields were dry enough to plant. He had Harlan on his way with a tractor and disk. Del unloaded the sheet metal in the machine shop and left for Prairie Point to get fertilizer. By nightfall they had seventy acres disked, fertilized, and ready to plant.

"Probably rain again tonight," Zach remarked with wry cynicism. If that happened, they could be putting in sixteen- to eighteen-hour days right through Memorial weekend, but he hated to have machinery on the roads during holidays. Too many impatient drivers—an exodus of city dwellers mostly traveling to the same places to become part of the congestion they sought to escape. If these periodic mass exoduses were attempts to renew rural roots, they were chasing illusions. Clichés of rural life were changing.

17

May 22
Spring Planting

By late May, spring already seemed exhausted, merging imperceptibly into the hot growing season of summer. There were more hours of daylight now with twilight lingering long into the evening. Grass was knee high in pastures, a sign that spring's transformation of the land was giving way to the fecund excesses of summer. Farmers, idled on the sidelines by too much rainfall, watched these changes with apprehension, worried as they saw a spring planting season slipping away, and waited for that uninterrupted string of fair-weather days that would dry their fields and let them return to the business of planting.

Then, with three days of drying weather, everyone was in their fields, and now even the ever-lengthening twilight couldn't add enough hours to the workday. Men on farm machinery swarmed the countryside, disking, fertilizing, planting, and moving tractors and farm equipment and supply trucks loaded with seeds and herbicides and fuel and water from field to field. They looked like farmers again, and felt like farmers, guiding powerful, dual-wheeled tractors that spewed thin curls of diesel skyward and dragged heavy tandem disks and harrows back and forth across long fields. From a distance, the machines seemed enslaved in a kind of slow-moving choreography, crawling back and forth across fields, scratching at the soil's surface, toiling over rain-hardened clods of dirt. In some cases, the farmers simply used no-till equipment and planted directly in the ground without any preparation. Despite high fuel prices, Del preferred to disk his ground. He reasoned that germination was faster in the loose soil because it warmed more quickly. Gaining even a day or two could make a difference, but he would plant no-till if time was short.

Del and Zach planted milo from dawn until after dark, until they couldn't see the line scratched in the soil by the planter's furrow marker, which showed them where to drive on the next pass through the field. Tomorrow they would finish the milo—310 acres of it—and switch to soybeans. It would take nearly a week to get their planned 600 acres of soybeans in the ground if all went well. Caitlin was home most of the time now. Work at the post office had slowed, and Amber had a permanent babysitter for Nathaniel, so she could help Del and Zach move trucks and equipment, or drive a lead truck with a "Wide Load" sign ahead of the planter and tractor on the highways. And there were snacks and meals to prepare at odd hours, especially when they worked late in the evening.

Some fields were still too soft to support the heavy fertilizer truck. The day before, Del had gotten his dual-axle fertilizer truck stuck three times in soft places in fields, managed to back it out of one muddy place, but twice had to drag it out backwards with a tractor. After that, he went to the fertilizer co-op and rented a smaller, twin-axle gravity spreader with big flotation tires that could be towed with a tractor. The co-op had a fertilizer-spreader truck with balloon tires and high clearance, and it would have been faster but someone had torn up the transmission. It would be down for weeks, too late for planting.

The next morning, Del finished breakfast before six and checked a small group of pregnant cows in a maternity lot. One cow was in labor but standing, and he managed to separate her in a corral. The tips of her unborn calf's two front hooves protruded slightly from her vagina, but the amniotic sac didn't appear to be broken. He made a mental note to recheck her mid-morning, then got an early start spreading fertilizer in the field across the road from the homeplace. When the spreader was empty, he drove the tractor, with spreader still attached, onto the blacktop highway and to the fertilizer co-op in Prairie Point for a refill. He didn't like driving tractors on highways because they were so much slower than cars, but the fertilizer plant was only a few miles away, not worth the time it took to unhitch the spreader and reconnect it to a truck.

At the co-op, Del discussed fertilizer strategies for his soybeans with Bernie Larousse, the manager. Bernie was graying and had a neatly cropped full beard. With his co-op cap slewed to the side and an affable demeanor, he greeted Del, and the two went inside. Del followed Bernie down a narrow hallway to an office in the back of the building. Bernie kept things running smoothly at the plant, but his office was in a state of perpetual entropy.

It was a lawless turmoil of papers and boxes piled on the floor, on his desk, and on every inch of shelf space behind the desk. Bernie motioned for Del to sit in an armless, metal folding chair in front of the desk. Del sat down, knees against the desk, back against the wall, and watched with amusement as Bernie turned his computer monitor and keyboard around to face forward. Del figured there was so much clutter back there that he couldn't get behind his desk. Instead, Bernie squeezed into another folding chair next to Del, pushed his cap back, and studied the monitor through the bottom of his silver-rimmed glasses. He suggested slightly altering the composition of the fertilizer, adding a little nitrogen. The soybeans didn't require nitrogen, but the cost was virtually the same with or without it.

"Did you have beans there last year?"

"No, it was milo ground," Del answered.

"A little nitrogen wouldn't hurt," Bernie continued. "Milo sucks a lot of nitrogen out of the ground. Might help your beans, especially since that ground hasn't been knifed with anhydrous."

Bernie tweaked the potassium and phosphorous components, too, ending up with a 7-40-40 ratio of nitrogen, phosphorus, and potassium. He punched a few figures into his computer, his thick, calloused fingers hitting the keyboard with more force than necessary, then jotted down the computed weights of the various fertilizer ingredients on a small notepad.

Outside, Del waited by his tractor and spreader while Bernie disappeared inside the bulk fertilizer building and climbed aboard a small yellow Bobcat loader with an unusually large scoop. Bernie scooted from bin to bin, deftly maneuvering the small machine, filling, backing, pivoting, dumping, then filling again, in a series of mechanical pirouettes as he loaded the large, rotating mixer. The dirt floor inside the building was wet, even muddy in places from the water-attracting nature of the fertilizer. The slightly acrid odor of ammonia given off by the fertilizer was strong enough to detect even outside. Del could smell it and was glad he'd waited outside. With the fertilizer components mixed, Bernie flipped a switch to start the auger. Almost immediately, fertilizer gushed from a pipe outside, and the spreader filled quickly.

As Del climbed into the cab of the tractor, Bernie looked up and asked, "You going to need that spreader all day? If not, let me know. Buddy Dean's waiting for one. Every spreader we own is in use."

"We'll need it all day," Del called out over the clatter of the tractor's diesel engine.

"That's fine. Just let me know."

"We couldn't get anhydrous ammonia on all of our ground. Some fields were too dry last fall, too wet this spring. And we've still got to spread phosphorus and potash anyway. Should be done by tomorrow though."

Midway through the morning, Del remembered the cow in labor, stopped his tractor and spreader at the edge of the field, and got into his black pickup truck. He drove three miles west on a blacktop county road, then a mile north on gravel, and turned onto the old Dunstan place at the top of a prominent hill. He continued past the abandoned house and stopped at a corral a little farther on. The cow, a black mixed-breed with a white face, stood wild-eyed and agitated in the corral. The waxy yellowish hooves of her as-yet-unborn calf still only projected a little, about the same as when he'd checked earlier in the morning. She hadn't made any progress except that her water had broken.

"Yeah, you're a problem," he muttered. "Looks like you're not having any contractions."

The cow stood her ground and watched Del get out of the truck. Climbing over the fence into the corral, Del spread his arms wide and started to herd the cow into a small alley in the corral where he could lasso and restrain her, but as he approached, she abruptly turned and charged, sending him running for the wooden fence. He paused just short of the fence, then advanced again, this time wary but still with arms spread, speaking softly to her. He had taken only a few steps when she charged again, this time sending him scrambling up the wooden corral fence for safety.

"Okay, if that's what you want," he said, muttering to himself and feeling a flush of annoyance as he headed back to the truck. "I can play this game too. I'll just tie you up right where you are."

The cow stayed by the fence, head up, agitated and ready to fight. Del pulled two hemp lassos stiff as old leather from behind the seat of the truck. He climbed partway up on the outside of the board fence, leaned out, and sent one lasso spinning toward the cow's head. In seconds, he'd wrapped it tight to the nearest wooden fence post and pulled hard to drag the cow closer. The cow pulled backward, but it was too late. Moments later, he readied the second lasso and slipped it around her neck and tied it for safety. With the cow secure and the ropes taut, Del dug into a heavy plastic box behind the seat of the pickup truck and located a long, clear plastic glove and his pulling equipment. He hoped everything he needed was there.

Wiggling his hand into the shoulder-length glove, he climbed back into the corral. Talking softly and resting a reassuring hand on the cow's rear end, he began the indelicate task of sliding his gloved hand inside the cow's vagina far enough to attach the ends of a small silver chain around each front foot of the unborn calf. With the use of only one hand inside the cow, it took a few minutes to get the two ends of the chain fastened. He could feel the calf's head pressed against its legs. Its head was in the correct position, directed rearward and in the same plane as the legs—a good sign. Without removing the long, plastic glove from his arm, he assembled the calf puller, an awkward-looking metal yoke that cradled the cow's rear end. He then threaded a pipe into the yoke with the pipe extending rearward. At the end of the pipe was a ratchet-like clamp called a "come-along." The crude design of the tool, which could have passed for a medieval torture device, belied its effectiveness. He'd saved the lives of many cows and calves with it over the years. In the hands of an inexperienced user, however, it could permanently injure a cow or calf or even kill them. Pulling a calf with it was one of Del's least-favorite jobs, a last resort. It was something he did only when he was sure the cow would die without assistance.

With the cow still standing and the shiny chain wrapped around her unborn calf's front legs, Del attached the chain to the come-along some three feet rearward. He then began extracting the calf in a manner not unlike using a jack and handle to raise a car. With each stroke of the come-along handle, the little chain tugged on the calf. Soon most of its dark forelegs appeared. Del paused, hoping the cow would push. She didn't.

"Come on, ol' gal," he pleaded. "Help me with this." The cow's posture remained unchanged. Del waited, then tightened the puller. Suddenly, the cow began to tremble, lost balance, and collapsed onto her side with the lassos holding her head up. With one final stroke from the come-along, the newborn calf slid smoothly, if unceremoniously, onto the bare ground, its tongue extended and its small, wet body glistening. Encased in its transparent amniotic sac, the calf looked like a lumpy, black fur coat gift-wrapped in wet cellophane. Its head moved. Del rubbed a hand across its nose to make sure its nasal passages were open and pulled the diaphanous membrane off its face. He stuck two fingers into its mouth to make sure its throat was clear of mucous, then rubbed the little calf with both hands to make sure it was alert and breathing. Its eyes blinked, and Del could feel its breath.

"Good," Del exclaimed as he climbed to safety on the other side of the fence. He slacked and removed the ropes constraining the cow, and almost immediately she stood up. Two feet of pinkish white afterbirth, like a blood-stained wet rag, hung limply from her rear. The cow lowered her head and pressed her nose to the calf. Del knew those first moments, when a cow got the scent of her calf, were critical if she was going to claim her newborn calf. "Now lick him and clean him up, and take care of him," he said, talking to the cow as if she would understand.

He gathered up his equipment, pulled off the still-wet and slimy plastic glove, and slung it on the pickup bed. As he turned the truck around and gunned it back toward the road, he could see the cow pushing on the calf and licking it, encouraging it to stand. Things would turn out all right.

It was a fine, warm morning. A scattering of high, wisplike clouds glazed a china-blue sky. In a brushy roadside ditch, a common yellowthroat, a tiny yellowish bird with a black domino hiding its face, sang a clear "witch-ity witch-ity" song, and paused to watch as a black pickup truck sped past, leaving a thin shroud of dust in its wake. Within twenty minutes, Del had washed up, disposed of the glove, and was back on his tractor.

Dinner was cold leftovers. Del was in a hurry. Zach called to say he would be finished planting milo by mid-afternoon on their eighty acres eleven miles to the west. He had enough milo seed but would need more herbicide. Two boxes of Bicep would be enough. He had a sandwich and thermos in the tractor cab and wasn't stopping for dinner.

Farmers often didn't apply herbicide when they planted because it slowed planting, having to stop and refill the planter's herbicide tanks with water and chemicals, and needing an extra trailer or truck with a water tank. Typically, a few weeks after planting, farmers sprayed their crops themselves or hired a commercial spray outfit. But if the weather turned wet and the sprayer rigs couldn't get in the fields, there would be a backlog of work with everybody wanting their fields sprayed at the same time, and somebody would end up at the back of the line. In today's high-stakes farming, a week's delay of a spray application could mean the difference between a profit and no profit. If fields were large, hiring a spray plane was always an option but expensive. Most farmers sprayed crops twice during the growing season. By applying herbicide when they planted, Del and Zach often needed to apply it only once more, and even if they never got back in the fields, the initial application would prevent a complete loss. Until a year ago, Del had never owned a high-capacity sprayer. Then he found a fully outfitted Ford truck, a high-wheeled unit with booms and tanks, and the price was right. It was an investment that would let him sleep better at night because he wouldn't be waiting in line at the co-op anymore.

It was true that herbicides were miracles of subjugation that helped American farmers achieve crop yields unmatched in the history of humankind. They could target specific plants, or affect plants in specific ways, and then disappear with scarcely a trace, all the while remaining relatively nontoxic to humans. But some were toxic to humans and wildlife, and there were financial and environmental and health issues that were, in part, the result of applying simplistic chemical solutions to complex environmental problems. The geometry of nature wasn't the perfect geometry of a field of soybeans planted in endless parallel lines. Nature was messy and complicated, and modern monocultures were littered with unintended consequences.

Before helping Zach move to another location, Del hooked a tandem-axle gravity feed trailer to his pickup and towed it to town, planning to pick up the first load of soybean seed for planting. If all went well, he could get the planter and sprayer refitted and cleaned and ready for soybeans planting by evening. He'd ordered the beans last winter, and on the way to town he called the co-op, which had sold him the seed, to let them know he was coming. When he arrived, he discovered they didn't have the beans in the large bags he'd ordered. To make matters worse, the seed company had changed the names on their seed lines, and the co-op wasn't sure which ones Del had actually ordered.

"Don't you have an old ticket somewhere from a previous sale where you can crosscheck names and numbers?" Del queried.

"No, not really," C.J., the manager, responded, shaking his head as he slid a pack of cigarettes back into his shirt pocket. "Everything's in the computer now, all switched over this past winter. The name of the variety doesn't show, only a stock number."

"You know, sometimes these dang computers make things worse than they were before."

Eventually, C.J. tracked down the correct name, only to discover that someone had ordered that variety in piddling fifty-pound bags—not the one-ton bags Del needed. C.J. was visibly upset. So was Del. It was as close as he ever came to swearing. He needed four thousand pounds of seed beans to get started and several times that amount to finish.

"Could they get more in a day or two?"

"We'll have a truck up here tomorrow, latest day after tomorrow with enough to fill your order," C.J. promised. "We've got enough here to keep you going a couple days anyway."

"Well, somebody better get themselves started opening about eighty of those bags out there now and load that hopper of mine."

Annoyed at the delay, Del unhooked the twin-axle trailer and instructed two of the employees to start opening bags and filling the hopper. He'd be back to get the seed beans later in the afternoon.

Caitlin rode with Del as they drove to the eighty where Zach had been planting milo. Eleven miles away, it was the most distant land they owned. With Caitlin's help, Del could get all of their machinery moved home in one trip—two trucks, a tractor, and a planter. Zach left in one of the trucks to help Harlan disk the next field. Caitlin drove the second truck with a wide-load sign, running interference on the highway for Del with the tractor and planter.

"Keep the lights flashing and the truck close to the center line so oncoming drivers will have to slow down," Del explained to Caitlin. "This planter's so wide that I can't get all of it on one side of the road."

With the tractor and planter back at the homeplace, Del began the changeover from planting milo to soybeans. He started by flushing the two 180-gallon spray tanks and the hoses and sprayer jets. Switching from one herbicide to another involved some down time to clean and flush every trace of a previously used herbicide from the system. Everything from tanks and hoses to applicator tips had to be spotless. Then he disassembled each of the fifteen sprayer jets and cleaned the tiny mesh filters in each one. He had used an atrazine-metolachlor mix called Bicep on the milo but would be using an older, less-potent pendimethalin on the soybean ground. The pendimethalin herbicide shipped under the name Prowl and came in a bright-yellow box with two two-and-a-half-gallon plastic containers of herbicide inside. The Prowl herbicide was also yellow and stained everything it touched, so after refilling the spray tanks with water, Del put on elbow-length rubber gloves before handling it.

With the herbicide tanks refilled, Del vacuumed out any remaining milo seeds in the long red hopper and reset the planting rate of each of the planter's fifteen drill rows. For milo, each planter tube dropped about four to six seeds per foot. That worked out to about five and a half pounds of milo seed per acre, or around 65,000 to 75,000 seeds per acre. Under average conditions, eighty to ninety percent would germinate. Soybean seeds, on the other hand, were larger than the buckshot-sized milo seeds but also had to be planted at a higher density because mature soybean plants were much smaller. To reconfigure the planter, Del verified the manufacturer's recommendations in the planter handbook. He had to get the planting rate correct or the entire crop would be compromised. Once underway, he would still need to monitor the planter's functions, but this planter—a relatively new one—had an electronic monitor in the tractor cab. The monitor was immensely helpful, keeping track of planting rates, depths, fertilizer application rates, and other items that affected planting. If a drill tube plugged, the monitor would signal the problem. Before on-board electronic monitors, a planter's seeding boot could plug, and a farmer might plant an acre or two before realizing there was a problem. Then it was hard to go back and replant just the skipped rows. Blank areas or skipped rows were once fairly common, and neighbors would always notice your mistakes. Such gaffs were rare today. The new planters were expensive but far more efficient than the old ones and assured accurate and uniform seed-planting rates. The newest ones were even GPS equipped. Feeding GPS data from soil tests and combine harvests of a previous year to on-board computers on their planters, farmers now had the ability to make tailor-made, on-the-go adjustments to fertilizer composition and seed density, which could maximize production in each field. Such systems weren't yet in widespread use by farmers in Del's area, largely because such computerized planting required extensive soil testing, and GPS-equipped machinery was expensive.

By six in the evening, the planter was cleaned and refitted for soybeans, and Del drove back to town to pick up the seed soybeans from the co-op. He knew, after the confusion earlier in the day, they'd have his seed ready. Twenty minutes later, he was back at the homeplace with his seed beans. He maneuvered the white, gravity-flow hopper close in behind the red planter, making it easy to distribute the soybeans evenly into the planter's hopper. The hopper was equipped with a special Archimedes'-type screw auger with soft, brush-tipped edges designed to reduce cracking or damaging seeds. Del also poured a layer of powdery black inoculant, a nitrogen-fixing bacteria, over the soybeans. It looked like coal dust. Del wasn't sure it made much of a difference. The bacteria were supposed to boost nitrogen production by infecting nodules on the roots of the beans, but the benefit was complicated.

Del squinted at one of the nearly two-dozen cardboard boxes containing the inoculation bags. He read the expiration date. It was almost a year out of date.

"Look at this!" he exclaimed to Caitlin, who'd come out to ask him about supper. "Last year the co-op did the same thing. Said the main office didn't ship new inoculants, so they gave me old inoculant. This stuff is a year old, too, and probably worthless just like last year. My guess is these bacteria are all dead."

"Well, you don't know for sure if they're all dead. Why not just go ahead and use it anyway? Can't hurt anything."

Del said he would but figured it was a waste of time. A lot of farmers didn't bother inoculating their seed beans. Still, it was recommended by the agricultural extension service, especially if soybeans hadn't been grown in a field for three or four years. Legumes like soybeans made their own nitrogen by extracting it from the atmosphere through a symbiotic relationship between the plant's rootlets and the bacteria that live in the little knots, called nodules, on the plant's roots. In theory, the inoculant was beneficial because it increased the number of bacteria available to "fix" nitrogen, and ultimately the flow of nitrogen to the plant. However, even without it, the beans would probably add as much nitrogen to the soil as they extracted. Also the bacteria were delicate, and inoculated seeds needed to be planted quickly, usually within half a day, or the bacteria would begin to die. And if the soil was too wet or too dry for several days after planting, the bacteria could die too. Maybe it helped, maybe not. Del wasn't sure.

One of the planter's hydraulic hoses that connected to the tractor's hydraulic pump wouldn't seat properly. Frustrated, Del hit it with a mallet, trying to force the hose coupling to connect. Instead he hit his thumb. Almost immediately, he saw blood oozing out from under his fingernail. He shook his thumb to get rid of the blood and shrugged off the pain. There wasn't time for bandages. It had been a long day, and he wasn't going to let anything as minor as a mashed thumbnail get in the way of planting this evening.

Caitlin had a sandwich and cold tea ready. Del went inside the house to eat, pacing in the kitchen, not taking time to sit down. She offered to get a Band-Aid for his thumb, but he said there wasn't time. Moments later, still carrying a piece of his sandwich in one hand, he headed for the door.

"Don't worry about me, hon," he said to Caitlin. "I'm going to plant until it's so dark I can't see the guide furrows anymore."

After he left, Amber stopped by after work to visit. She and Nathaniel followed Caitlin as she watered flowers and toured the garden, all to Nathaniel's delight. He was two now and trying to touch everything—leaves, flowers, bugs, the big hollyhock plants. With the men and their noisy machinery gone, the homeplace seemed unnaturally quiet, the evening serenity broken only by the unhurried phrases of a distant mockingbird and the mellow chatter of purple martins as they swooped overhead and fluttered in front of their small apartment quarters overlooking the garden.

"See what kind of life you married into?" Caitlin said. "Del just loves this sort of thing, and so does Zach. The busier they are and the harder they work, the better they like it. But there are times, like spring planting and harvests and sometimes haying, when you won't see much of them. I just try to keep them fed and help when I can."

Amber said she was beginning to understand. Not growing up on a farm, she hadn't understood the intensely seasonal nature of farm work and just how much farming success depended on getting critical work done at the right time. There would always be work to do on a farm, but some of it was more urgent than others. Farmers knew that the march of time was both friend and foe.

18

May 28
Miracle Seeds?

The soybeans that Del and Zach were planting looked just like the ones they had always planted—shiny, round, pea-sized, and bullet hard. The new crop, maturing three or four months from now, would look the same too. But it was an illusion. Hidden deep in the bean's genetic code, in its DNA, which determines what each living organism becomes, was an important difference. These soybeans contained a gene that made them resistant to an herbicide lethal to other plants. It was only a single implanted gene, but it had ignited a firestorm of controversy. Why would a trait so valuable—something that would reduce the amount of herbicide used, lower production costs, and increase yields—find itself in the middle of a controversy?

The lowly soybean seems an unlikely candidate for controversy. Soybeans belong to the *Fabaceae* or bean family, one of the largest plant families and one of the most valuable to humans. Soybeans first appeared in human history in China more than five thousand years ago. Hundreds of varieties were developed but remained unknown outside of China until they appeared in Europe in the 1700s. They reached North America in 1765 when a Georgia farmer named Samuel Bowen planted them and produced a soy sauce. He eventually exported more than a thousand quarts of soy sauce to England, but his success remained modest. James Mease extolled the virtues of soy sauce in the 1804 edition of *The Domestic Encyclopaedia*, saying "it is chiefly used at the tables of the luxurious," and added that soybeans grew well in Pennsylvania and ought to be cultivated. But few heeded his comments. Later still, soybeans were recommended for livestock forage and as a nitrogen-builder in soil. In 1896, an article by Henry Trimble titled *Recent Literature on the Soja Bean* portrayed soybeans as a multi-purpose food for humans, mentioning soy sprouts, soy oil, miso, natto, tofu, soy sauce, fresh green soybeans, even soy coffee.

Even in the early twentieth century, soybeans were still grown in the United States mainly as a forage crop in the south. Finally, in the 1940s, with advances in food technology, soybean seed production expanded into the Midwest, and the United States quickly became the world's largest producer and exporter of them.

The ascendancy of soybeans as a world commodity in many ways parallels the rise of science knowledge. Selective breeding increased its protein content from twenty-eight to forty percent, and its oil content was increased to more than twenty percent. The oil, in particular, made soybeans valuable as fat emulsifiers in foods and for industrial uses in varnishes, caulking compounds, linoleums, inks, paints, even lubricants and diesel fuel substitutes.

Given such a long history of cultivation, it might seem surprising that the soybeans Del and Zach were planting were generating so much controversy. However, they were planting Roundup Ready soybeans, a heavily patented seed type not available commercially prior to 1996. These soybeans were one of a number of new products generically known as GMOs, or genetically modified organisms in the parlance of the literati. They differed from the ones Del had planted previously in being immune to the effects of the herbicide Roundup, a glyphosate-based chemical toxic to broadleafed plants. The value of such a genetic blueprint was obvious because the young bean plants were not harmed by spraying to control weeds and grass. Farmers quickly recognized the economic potential of Roundup Ready soybeans and of other GMOs.

The GMO controversy was partly over how the genetic modifications were made and partly over their impact on the environment. Genetic modification through selective breeding has taken place for thousands of years, but the changes were gradual and involved characters naturally present in the plant's gene pool. On the other hand, altering the genetic makeup of a living organism by physically inserting new genes—tiny pieces of DNA—into the chromosomes of the plant's cells was new. To achieve the desired result, the new genes could even come from unrelated organisms—from, for example, a bacteria *Bacillus thuringiensis* (Bt for short), a different plant, even a fish. Plant varieties achieved this way are known as transgenic crops. The goal of transgenics is the same as that of any selective breeding program—to achieve favorable traits—but transgenics expand possibilities beyond what could be possible with traditional cross-pollination and selection techniques. Most alarming to environmentalists, it expanded these possibilities almost overnight. The results of transgenics were immediate and dramatic and the consequences, in some cases, worrisome.

American farmers initially embraced transgenic technology. Most controversial, however, were the Bt transgenics—Bt corn, Bt potatoes, Bt rice, and so on—which now contained in every cell of the plant a bacterial gene producing an insecticide lethal to insects. Because the insecticide is also carried along in the cells of the seeds, the seeds also become a potential hazard to whoever eats them. Critics proclaimed that it was only a matter of time until the two kinds of seeds, Bt and non-Bt, got mixed-up on a farm or in an elevator. And, when it did happen with a transgenic corn, headlines screamed the fears of critics. "Seeds of Disaster," one said; "Seeds of Conflict," another said; "Harvest of Fear"; "Biodevastation"; "Farmaggeddon"; "Frankenfood" and so on. Cries for boycotting transgenic foods, especially in Europe, brought transgenic projects to a screeching halt. Then multinational seed companies with large amounts of money invested in GMO and transgenic research, adding fuel to the fire by slapping patents on their genetic products to protect their investments. Patents made it illegal to save back some of the crops as seed for replanting, a time-honored tradition, especially in poorer nations. Further, transgenic crops were developed that produced seeds that would not germinate—essentially eliminating the possibility of any replanting from harvested seeds. Farmers began having second thoughts too. Nobody wanted to grow something they couldn't sell.

Del knew that Bt transgenics were a hazard. He had never planted any of them, but he felt that opposition to all transgenics was unfounded. Transgenics, especially self-pollinating ones like soybeans where there was little danger of the new genes becoming established in unwanted areas, had the potential to improve yields, improve resistance to drought, and improve resistance to saline soils. These and possibly other traits could benefit farmers, as had Roundup Ready soybeans.

A few years back, at the height of the controversy, a college student interviewed Del and Zach about biogenetic issues. She'd come all the way from California for the opportunity to hear first-hand what some midwestern farmers thought about GMOs and transgenic technology.

Del told her that he never had clean fields before and usually had to apply a lot more herbicide to control weeds. With Roundup Ready soybeans, he applied a pre-emergent during planting and later sprayed with a single application of Roundup. It was a big savings economically. Clean fields meant better yields, and he didn't have the expensive, labor-intensive cultivation to control weeds like he used to have to do. And because he didn't have to cultivate anymore, there was much less soil erosion and less soil compaction, and soil moisture was conserved. He also said that soil scientists told him he probably also conserved soil microflora and fauna this way, but he didn't know about that for sure.

When she asked about the risk of transgenic gene flow into related plants bordering the fields, or the effects of Bt transgenics on wildlife such as butterflies and birds, Del said he didn't really know. He pointed out that it was important to compare the risks of transgenics with the risks of conventional herbicides and pesticides. There were risks with both. He said environmentalists needed to understand that, by and large, some kind of chemical control probably would always be needed if large-scale, monoculture farming was to continue. He also thought it would be large-scale farming, not small organic and designer produce farms, that would feed most of the world in the future.

Del also told her that in farming it wasn't a matter of choosing one kind of control or none. You had to choose one or the other. Most of the transgenics looked a lot better in that kind of comparison.

When the student asked Del what farmers did to control weeds before herbicides were used, he said farmers used row-crop cultivators that tilled the soil. He said that farmers used to spend half their summers cultivating corn and beans. Besides being frustratingly slow, it exposed loose topsoil to a lot of erosion. He stressed that anybody who thinks GMOs only benefit big corporate agriculture had never been on a smaller farm. Family farms have also benefited a lot from GMO technology, even if the seeds were expensive.

Del and Zach were getting a break in the weather and hoped to finish planting in two days, maybe three. Harlan, driving a dual-wheeled Farmall 1085 tractor, was disking from early morning until six in the evening, sometimes later. The twenty-two-foot tandem disk he used sliced the soft, damp earth with each pass and left a pliable seedbed in its wake.

Even so, Harlan couldn't disk ground fast enough to stay ahead of the planter, so after he went home in the evening Del took over, disking late

into the night to keep a seedbed prepared ahead of the planter. With only occasional stops to refill the planter's herbicide and seed tanks, Zach's ground-eating pace was impressive. He also planted until well after dark, not stopping until after 9 p.m. when it got so dark he couldn't see the planter furrow markers in the tractor's lights. Del told him that if they continued to expand their farming operation, they'd have to upgrade to larger tandem disks and bigger tractors. That was his goal before the next planting season.

During the day, Del alternated between spreading fertilizer, keeping a supply truck loaded with seed beans, a water tank filled for the herbicide mix, and the trucks and trailers moved up into position close to where Zach was planting. When time permitted, he disked alongside Harlan with another tractor and disk. It was a lot to juggle. If he fell behind, planting was delayed. To keep things running smoothly, Del and Zach kept in constant phone contact, Zach advising Del when he would need seed soybeans, or water, or herbicide, or when a truck needed to be moved closer. Del thought it was ironic that farmers never seemed to use the term "multitasking," but it could have been invented right here on the farm.

Del, like many midwestern farmers, hadn't used a moldboard plough in years. Seedbeds were either planted using a no-till planter or prepared ahead of time with a single pass of a disk. Crude ploughs of various forms had been used in agriculture almost since recorded history, and there had been little advance in design until American blacksmiths in the early 1800s tinkered with new shapes and the use of iron instead of wood. In 1837, John Deere marketed the first cast steel plough. It's self-polishing surface readily cut through the soil and made it possible to break tough tallgrass prairie sod. But today these wedge-shaped moldboard ploughs were almost anachronisms, implements with little or no role in modern American agriculture; using them was time-consuming and they were a major cause of soil erosion. Since the 1970s, farmers had abandoned them en masse for tandem disks, or tandem disk harrows, which could prepare a seedbed much faster and were less invasive to the soil.

Now another revolution was underway. Many in agriculture today were espousing the benefits of no-till farming, a throwback to agriculture before modern farm machinery. No-till planting was just what it sounded like—planting with no seedbed preparation. Heavy specialized planters cut slits in the soil, dropped seeds in the slits, and pressed them shut, all in a single pass at four to five miles per hour, faster than a person could walk. No-till was gaining acceptance because it lowered planting costs and minimized erosion. On the negative side, undisturbed soil didn't warm or dry as quickly as tilled soil, sometimes making it difficult to get seeds planted in a timely fashion. Del was ambivalent about no-till techniques because germination could be delayed or lowered if soils stayed wet and cool, and that was an issue especially with early spring corn planting.

Del thought no-till worked better for soybeans, which were planted later in the spring when soils were warmer, than it did for corn. Del had found that no-till planting was most useful for double-cropping, such as planting soybeans immediately after a wheat harvest when you were in a hurry. But there were drawbacks. No-till took more herbicide, or more expensive herbicides, which partly negated savings in fuel costs by not disking. Proponents argued that no-till soils became mellower with age, and university research had shown that, over time, no-till soils gained in organic matter and carbon, a benefit on both accounts. Today Zach was using a no-till planter capable of planting in soil without any seedbed preparation, but, if they had time, he and Del still preferred to prepare a seedbed with a tandem disk first.

Del had helped Harlan disk this morning because the bottomland field they were disking backed up against a narrow wooded stream and was wetter than their other fields. Slippery elm, hackberry, black cherry, and Osage orange lined the streamside. A few years before, a family of beavers had taken up residence on the stream, and their dam flooded several acres of the adjacent field. J. P. Henderson, the landowner, was a tall, lanky military veteran, and he took the beavers' engineering efforts as a personal affront to his land stewardship. He dynamited the dam and once waited by the creekside for several hours hoping to get a shot at them, but the beavers never showed. This year the beaver dam was gone, and the field, once covered by a shallow pond, was now invaded by a weedy carpet of smartweed and cocklebur seedlings.

"Looks like them beavers are gone," Harlan said, using his cellphone to call Del from the cab of his tractor on the opposite side of the field, his voice breaking up against the clattering background of engine and tractor noise. He could see that Del's tractor was headed toward the wet area of the old beaver pond.

"I guess so," Del responded. "I checked over the fence and didn't see any sign of a fresh dam or of them chewing on trees."

"Think that low area is dry enough for you to disk it?" Harlan asked.

"Maybe. Guess I'll go see if I can get stuck in it," Del added with a hint of sarcasm. The damp, fresh-cut soil behind Del's tandem disk left a band of dark, aromatic soil in its wake. Whenever soil was ploughed or disked, there was always a special fragrance associated with it that was caused by the release of carbon and organic compounds bound up in the soil. It was particularly strong in the damp soil this morning, and he savored the rich, earthy aroma that even permeated the air inside the insulated tractor cab.

Zach continued to plant soybeans in a field on the east side of the road, opposite where Del and Harlan disked. Del's blue 1967 Chevy truck with a gooseneck trailer attached was parked on a grassy strip at the side of the field. On the trailer was a white, thousand-gallon water tank and dozens of yellow boxes of Prowl herbicide. Behind the truck and trailer was a black flatbed pickup hitched to a twin-axle trailer with a gravity-feed hopper full of seed soybeans. By late morning, Zach needed to move to an adjacent smaller field in an inconvenient location that required crossing a small creek. Not wanting to walk back and get the support truck just for that field, he called Del.

"I got about sixty-five gallons of herbicide left in each tank. Is that enough to plant the corner field across the creek or do I need to fill the tanks first? I hate to cross the creek with the planter fully loaded if I don't have to."

"Hold it a minute," Del said, speaking loudly into the phone as he stopped the tractor and idled the engine. "Can't hear you. Okay, tell me again how much herbicide you got left in the planter tanks?"

"About sixty-five gallons."

"And how many acres you want to plant?"

"About fifteen."

Del pulled out a small spiral notebook and laid it on his knee. He scribbled a few figures. "Looks like to me if you add another sixty-five gallons to each tank and split a two-and-a-half-gallon container of herbicide between the tanks, you'll have enough for twenty acres."

He knew Zach would always call and confirm little calculations like that. It was as if he didn't completely trust himself even when he knew the answer.

"Thanks, Dad," Zach said, then added, "I flushed a wild turkey off a nest this morning. It's along the north fence, about six posts back from the low-water crossing on the Henderson place. She's got seven eggs."

"I'd seen the ol' Tom over near there myself early this morning," Del replied. "All puffed up, tail spread and ever'thing. Lots of turkeys up and down that creek. Try not to get too close to the nest with the tractor."

"By the way, what're we doing for dinner?"

"Caitlin's bringing fish sandwiches from Long John's. We can eat under the tree by Henderson's driveway. It's close. Save us time that way. I'll let you and Harlan know when she gets here."

It was nearly one p.m. when Caitlin drove her blue pickup under the tree at the edge of the grassy driveway less than a quarter mile from where the men were working and laid out sandwiches and drinks on the tailgate. It was a hot day, more like midsummer with a light breeze, pleasant in the shade. The four of them gathered to eat and rest—a brief convergence of people and machines. Harlan flopped down on the grass and leaned against a truck tire to eat his sandwich. Del sat in the opened doorway of the truck, his back to the big red Magnum tractor and disk parked behind. Zach looked up at a towering column of poison ivy vines climbing the shade tree overhead. He'd gotten it a couple of times and decided to sit on the pickup tailgate, as far from the poison ivy as possible. Caitlin stood by the tailgate, using it as a table. The men talked, worrying about soil moisture, and rain, and seeds, and fertilizers, and the mechanical condition of their equipment, and about how to move their trucks and machinery most efficiently from one field to another. Zach said he was getting low on soybeans. Del said he'd get more seed beans from the co-op after dinner. They talked and ate and enjoyed a moment of relaxation, a respite from unrelenting work. Behind them a pair of bluebirds chattered softly, then fluttered in front of the curved exhaust opening of Del's Magnum tractor. The exhaust was hot, and the birds couldn't alight for more than a second or two in its opening. Undeterred, they kept returning, hovering then fluttering away.

"Look at those birds," Caitlin exclaimed. "What are they doing there? Trying to catch bugs?"

Del turned around to look at the birds, then laughed. "I don't think so. It's a hole. They're looking for a nest hole. They got a hot one there, too, and it ain't gonna be there much longer neither." Their break lasted little more than fifteen minutes.

After the meal, Del drove to Prairie Point to get seed soybeans at the co-op. It was easy now. No handling of hundreds of little fifty-pound bags like when he first started farming. And this time, after the mix-up at the co-op a few days ago, they'd have the soybean seed in bulk. Del parked the truck and gravity-flow trailer crosswise at the front of the feed company's huge Quonset building, and Dwayne, one of the elevator workers, yelled to Del, "What variety you want? Denver or Dallas?"

"Dallas," Del shouted over the elevator noise. He wondered why a seed company would choose to name their seed varieties after cities.

Dwayne, wearing a heavily soiled cap and a plaid shirt with the sleeves ripped out at the shoulders, nodded. He'd been pretty sure what variety Del wanted even before he told him. Dwayne spun the forklift around and hooked a giant canvas bag filled with 2000 pounds of seed soybeans. Pivoting the lift back around, he raised the huge bag high up over the bright-orange hopper, the lift rocking a little with the heavy weight suspended overhead. Joe, a younger and newer employee clad in jeans and yellow t-shirt, scrambled up the ladder on the side of the hopper, then leaped to the top and lightly balanced himself with the confident casualness of youth some eight feet up, his feet positioned on either side of a corner of the big hopper. With the forklift holding the huge tan-colored bag suspended over the hopper, Joe leaned toward it slightly, held onto the bag with one hand for balance, and with the other untied a drawstring pouch at the bottom of the seed bag. In seconds a ton of seed soybeans plummeted into the hopper. The forklift recoiled visibly, and the hopper shuddered as a dust cloud rose up from it. Joe, unperturbed, remained balanced atop the hopper while Dwayne hoisted up a second one-ton bag. The two bags Del needed—four thousand pounds of seed beans—were loaded in minutes.

"Keep track of this," he said to Dwayne as he turned to leave, looking down momentarily to step over a dead mouse lying in the dust at the edge of the wide concrete doorway. "I'll get a ticket for these later."

Caitlin helped Del track the mounting expenses through the two-month-long planting season. They would go over the burgeoning stack of receipts together, checking entries in their farm records book and later, on the computer, trying to recall each expense as it was entered. Errant expenses like the one Del had just left at the co-op when he loaded the gravity flow wagon with soybean seed were the hardest to remember. It was no small task, and expenses were increasing every year. Even with volume discounts, hybrid seed corn cost 140 to 150 dollars a bushel, and that would plant only about three acres, figuring around 40,000 seeds per acre. Their seed corn alone exceeded $30,000. Roundup Ready soybean seed was less expensive, about fifty-five dollars a bushel, but farmers planted three times as many soybean seeds per acre as corn, up to 150,000 seeds or more, so the planting cost was about the same. Del would also spend over $30,000 on soybean seed this spring. Adding the heavy 120-60-60 fertilizer, some less-expensive 7-40-40 fertilize he'd used earlier, the anhydrous ammonia, and the herbicides brought planting costs for both corn and soybeans to about $250 per acre. His costs would amount to well over $300,000 for these two crops, and that didn't including the 310 acres of milo he planted. And the crops would still need an additional application of herbicide. If he planted wheat, which he did some years, there would be still more costs. And the fescue had to be winter fertilized too. And then there was labor, crop insurance, the inevitable machinery maintenance, and cash rent due on rented land. Diesel fuel for their fleet of tractors and trucks could add another $6000-9000 a month, a cost that would continue through the end of fall harvest in October or early November. Those were just the major items. Caitlin watched the figures rise and sometimes got a sick feeling in her stomach. What if something went wrong? Or somebody got hurt?

Would it rain enough? There were so many things that could go wrong. No wonder farmers worried about the weather and commodity prices and crop diseases and weeds and just about everything else.

Del worried, too, but tried to make light of it. "It's about the same as it used to be, just more zeros now," he said to Caitlin.

"Well, it's a lot more zeros now than when we first started," she said. Much of the year's success or failure would be riding on the spring planting season. Del knew a spring or summer storm could wipe out a season's work. So could a drop in commodity prices or an insect infestation. That was why he bought crop insurance. It was expensive, but when farming at this level it was a necessity. It put a floor under losses and let farmers stay in the game. Sometimes he also hedged some of his crops in the commodity market, but that meant at the end of the season he had to produce the crop. Del often said, "No matter what size your farm operation, you just hope there's more good years than bad ones."

The Montgomerys pushed hard until spring planting was finished, not wasting a minute, eating in the fields, working after dark until they couldn't see planter marker lines in the field even with the tractor's lights, making repairs at night if necessary, and rising before dawn the next morning to start over again. Even Harlan worked extra hours. They worked doggedly, driven to seize the moment because they didn't know if another would come. The weather offered only so many windows of opportunity, and losing even a day or two could cost tens of thousands of dollars. If the weather remained favorable, they could finish in ten, maybe eleven days. Last thing they wanted was another rain delay that could push planting into the first week of June, like last year, when it overlapped fescue harvest. They would need to start swathing fescue in a week to ten days. More rain, both provider and plague to farmer, was forecast for Sunday.

19

June 11
Fields of Fescue

June. Rain-delayed spring planting may continue into early June; farmers begin swathing fescue grass for seed; fescue harvest underway by middle of month; bulls placed with cowherds for summer breeding; farmers apply herbicide to early planted crops; tractor pull competitions held on Saturday nights; wheat harvest begins late in month; soybeans double-cropped with no-till planters after wheat harvest if time permits; orange butterfly weed brightens roadsides; ripe cherries attract birds; mulberries ripen.

Field corn was waist high and corn leaves were whispering in long rows. Soybeans with pairs of freshly sprouted, ankle-high leaves dissected black fields with razor-thin, green lines that converged on distant horizons. And all across the countryside, bright fields of fescue grass higher than the bellies of cows swayed in afternoon breezes. Now the sun was passing high across the sky, overseeing the transformation of spring's youthful vigor into summer's maturity. Some said that at this time of year it was possible to hear the corn and other crops growing, and it was like a symphony. Each evening as darkness settled across the land and mighty Arcturus, the brightest star in the northern sky, faithfully climbed to its place overhead, a symphony of another kind could be heard. That symphony, the piping of myriad crickets and frogs and frequent refrains from a love-struck mockingbird or the cry of a night bird, was played out beneath a vast celestial stage. For a moment on those perfect June evenings, everything on that stage—the fields, the sounds, the stars, the night air redolent with the scent of fresh-cut grass—seemed eternal, as precisely tuned as time itself. It wasn't, of course, because just as symphonies have beginnings and endings so do seasons. Even brilliant Arcturus wouldn't escape the cycle. On its steeply inclined passage obliquely down through the great disk of the Milky Way it would, one day a half-million years from now, fade from sight, too, leaving the Great Bear constellation alone to roam the northern sky.

With the summer growing season underway, there was now a steady flow of farm work. For the first ten days in early June, maturing fescue seeds were past boot stage but not yet mature. This worried farmers. The fescue seed crop was vulnerable to a summer storm. Half a crop, even more, could be lost to seed shattered on the ground in a single rainstorm. This was swathing time, and Del and Zach, like their neighbors, would be hitching swathers to tractors and checking belts and tensions, greasing fittings, and pouring oil on long cutting sickles. Maybe they'd knock a few mud dauber nests off the machines, too, and they'd listen for squeaks and rubbing sounds that could become problems.

A swather, like a lot of agricultural machinery, is a straightforward machine. It has only one function—to cut a hay or seed crop and arrange it in a swath or windrow so it can mature and dry more quickly. Once in a swath, there is less chance of losing the seed to a storm. Swathers cut grain the same way combines and sickle mowers do—scissoring it with a menacing row of trapezoidal-shaped steel blades called sections, which are riveted to a flat bar that slides back and forth between conical steel guards that point forward. Idling, a sickle clatters noisily. At normal speed, the sickle's motion is a terrifying blur—like a twenty-five-foot electric knife—and unnerving when one begins to imagine what it could do to careless fingers or hands. As a swather is pulled forward through a field of fescue or grain, a large, rotating paddle-like reel, similar to those on a combine, gently pushes amputated stems with seed-laden heads onto two broad, flat canvases. The canvases, like moving sidewalks, transport the prostrate crop from each end toward a central gap, and there the cut stems tumble down onto the stubble below and form a ropelike windrow with stems overlapping like tipped-over dominoes. It happens fast, tractor and swather moving six or seven miles an hour, sickle-bar slicing back and forth a thousand strokes a minute, its chattering teeth full of fescue. In the blink of an eye, a crop shorn of its reproductive germplasm is transformed into a gigantic, spiraling maze of windrows like a forty-acre doodle from a bird's-eye view. To a farmer it resembles nothing so much as a crop one step closer to money in the bank.

Mechanically, a swather is relatively trouble-free. A generation ago many were self-propelled machines that steered from the rear by working two leavers up front where the operator sat. Because of the rear steering, these machines always looked like they were moving backwards through fields. Today most swathers are larger and pulled by tractors. The work of swathing is repetitive. It demands little except staying alert. While they drive, farmers spend half their time looking backward over their right shoulder, watching the sickle and the canvases and the steady, conveyer-belt flow of cut stems being piled into a fluffy row. Trouble seldom arrives until late evening when grass stems become damp and tough. When that happens, the grass stems wrap around the long rubber rollers that turn the canvases. Once wrapping starts, it worsens until something breaks or overheats, and prudence dictates a low-tech solution. Stop, climb down from the tractor, and cut away the wrapped stems with a pocketknife. That's when farmers know the end of a long day is near. On dry summer evenings that might not happen until well into the night.

The two big swathers that Del and Zach owned were heavy, unwieldy-looking machines, but they did their job extremely well. With each pass a twenty-five-foot span of crop was felled, the razor edges of the sickle slicing through fields like a plague of locusts gorging on fresh grass. Zach, driving

a red International tractor and pulling a red swather with a silver batting reel that flashed in the sun as it turned, cruised steadily around the field, transforming an ocean of silvery green fescue grass into a close-cropped landscape of windrows. An occasional rabbit or cotton rat scurried ahead, zigzagging in panic at the swather's chattering approach. In the swather's backmow, a banquet of disturbed insects rose up, taking to the air to flee the disturbance, and were assailed by a platoon of swallows tirelessly dipping and circling behind man and machine.

This was the fifth day of swathing, and Del and Zach figured they could finish in three or four more days if the weather remained dry. They could have swathed more acres on shares but declined several jobs because field conditions were uncertain.

Desperate for help, one neighbor said, "Well, there are some sprouts out there, but if you want I can chop them out by hand."

Del shook his head and tried to look sympathetic. "To tell you the truth, I don't think we can get to it in time."

Del knew woody sprouts could break a sickle or a guard on the cutter bar or bend the batting reel of the swather. A cut sprout was worse. It could puncture a six-hundred-dollar tractor tire. It wasn't worth the risk. He wasn't going to do a job another guy didn't want to do and risk his machinery and an expensive breakdown.

Fescue seed wasn't the largest cash crop on the farm, but it was important, and Del was constantly preoccupied with contingency plans, trying to prepare for anything that might slow up swathing or the harvest that followed. Fescue seed prices had hovered mostly between twenty and thirty cents a pound for forty years. Occasionally prices spiked higher, even doubling when yields were low and farmers didn't have as much to sell. Like a lot of things in farming, low profit margins could be offset by high volume, but it was a vicious spiral that kept production high and prices low. There was little farmers could do to break the cycle, so they continued to ramp up production. The economics were straightforward, but there were minefields on the way to the goal.

Sometimes Harlan, the hired man, was one of those minefields. He'd showed for work almost every day this summer, but it hadn't always been that way. A year ago, at the onset of swathing, he abruptly announced that he'd entered a crappie fishing tournament. He left for a week. Then two weeks after returning, he entered another in Arkansas, leaving them short-handed during harvest as well. Del was exasperated. Caitlin helped until they found a friend of Zach's who was able to help truck seed to elevators in the evenings and run a combine. Harlan's erratic absences were an inconvenience, but Del was willing to overlook them—to a point—because skilled farm labor was difficult to find. Despite Harlan's late arrivals, fishing tournaments, and assorted crises that stalked him, he was a seasoned farmhand, experienced with cattle, and skilled at operating complex and increasingly expensive farm equipment. Also, he could recognize mechanical problems when they arose and often do the repair work unassisted. It wasn't a job description for inexperienced seasonal laborers or idealists hoping to spend a few idyllic weeks on a farm during the summer.

Harlan and his fishing buddy hadn't fared well in either tournament, and the four-hundred-dollar entry fees they split were expensive for them.

When he returned, Del said, "Harlan, you ought to know you can't compete with the professionals in those tournaments. They have lots more time and money than you do for that sort of thing."

"I know it," Harlan agreed, but he seemed unfazed by his lack of success. Then, a bit defensively, he added, "But we got ninth place in that Arkansas tournament."

Del conceded that Harlan's showing wasn't bad considering there were nearly a hundred entries, and Harlan had returned to work bubbling with enthusiasm and talking about entering again next year. The fact that the tournament might interfere with work didn't seem to occur to him. There would be three categories of prizes, he enthused, his eyes brightening at the possibility of winning them all.

"A fourteen-hundred-dollar purse. Can you imagine?" he exclaimed. "I could sure use that money."

Del was pleased that Harlan had clocked forty-seven hours this week and was on time every day. Caitlin made out his weekly paycheck while Del ate breakfast. "This is the first time this year he's managed more than forty hours in a week," she said.

"Probably true," Del added. "Even if he works late, he hasn't been able to get in forty hours because he oversleeps and gets here so late."

Fescue is a grass. Like many grasses, it's in the select company of plants that are among the world's most valuable to humans. Some varieties of fescue are native to North America, others apparently to Europe, but their true origins are murky. Like all grasses, fescue has a jointed stem, which is solid at the joints and hollow between, but identifying varieties of grasses, as anybody who has studied them knows, requires navigating a microcosm of arcane vocabulary and structural floral minutia daunting even to professional botanists. Julian Steyermark's authoritative *Flora of Missouri* weighs in with eight species in the genus *Festuca* and numerous varieties; two species and five varieties mentioned are of European origin. That was in 1963. Today there are even more varieties.

The fescue the Montgomerys were swathing was K-31, a well-established variety first harvested in Kentucky in 1931. By the 1950s, other varieties had been developed. All of them are hardy grasses that remain green year-round but grow best during cool spring and fall periods. Fescue is tolerant of drought and poor soil, which makes it popular for erosion control along highways, in lawn-seed mixes, for pastures, and as a hay crop. It thrives almost everywhere except the South.

Originally, fescue was a bit player on the world's stage of grasses, but plant genetics, fertilizers, and human-assisted introductions spread it far and wide. Looking at a field of fescue, you wouldn't notice anything remarkable about it. It develops a dense, dark-green sod, and the stems are tall and slender. Seeds are born in spikelets at the top of the stem, but they are small and light in weight, not large and heavy like wheat or oats. For top yields, fescue needs hefty boosts of nitrogen. Its wide distribution across North America belies the fact that, as a seed crop, it is grown mainly in just three areas: Oregon, southeastern Washington, and western Missouri. In the northwest, it is grown for seed and yields a ton or more of seed per acre. Missouri farmers squeeze a seed crop, summer hay crop, and late-year pasturage from a single field but sacrifice seed yield for the accrued advantages of a hay crop and cool-weather pasture. A few midwestern growers forgo the seed crop in favor of an early hay crop when plant nutrition peaks and hay buyers pay a premium price for early-baled fescue.

Del and Zach started swathing fescue for seed at the end of the first week in June, when the fescue was green and moist. Ten days and thirteen hundred acres later, when they finished, the early-swathed seed would be dry and ready for harvest, so there would be no break in their work. Then they would bring two of their three combines out of storage sheds and attach small, stripped-down combine headers. On each header there was a pick-up attachment—a flat platform consisting of two wide rubber conveyor belts—mounted out front. The belts bristled with springy steel tines that deftly plucked up the windrows of dry fescue and guided them, like big, loose ropes of grass, into the throat of the combine. With an application of fertilizer in late fall or winter, a Missouri farmer might expect four hundred to six hundred pounds of fescue grass seed an acre in a good year, half that or less in a dry year. Del hoped he could net a hundred dollars an acre. He remarked that nobody would get rich growing fescue, but it helped pay bills and made a little hay and decent winter pasture for the cattle. But first they had to finish the swathing.

Del, Zach, and Harlan kept the two swathers running nonstop sunup to sundown. Del left the house before six every morning and operated one of the swathers until eight-thirty when Harlan was supposed to arrive. He'd take over again briefly at noon while Harlan ate, and again in the evening after Harlan left, usually staying in the fields until dark or later. Only fatigue and damp evenings when the grass wrapped on the swather's rollers would prevent them from working later. Zach worked the longest days, keeping his own schedule, eating a midday sandwich on the go in the cab of the tractor. Amber usually took supper to him in the fields after she got home from work, and she'd bring Nathaniel along so he could ride in the tractor cab with his dad. Inside the cab, Zach put Nathaniel on his knee where he was next to the console with the shift lever. Nathaniel loved riding there, but most evenings he fell asleep soon after the tractor was underway. One evening, in an unfamiliar field, Zach worried that he might hit some old ruts or a rock. "It's too dangerous," he said to Amber, "I can't watch him and the equipment at the same time. Just let him watch for a while."

It was a decision he almost regretted, precipitating a crisis of misunderstanding for a two-year-old.

In between running the swather when Harlan wasn't around, and keeping the tractors fueled and the swathers greased, Del ran errands and occasionally checked his cowherds. A few holdout cows still hadn't had calves. The night before he had noticed one that was ready to calve. She was alone, away from the herd, her udder looking swollen and her vulva enlarged and with mucus discharge. This morning she was attending a still-wet, moist-eyed calf. Del picked up the calf, a heifer, and set it on the back of his truck, then gave it, just hours old, its coming-of-age party—a vaccination and an identifying ear tag. The ear-tag punch hit a vein and blood squirted, dousing Del's pant leg and dripping in a stream down the calf's jaw. He ignored the bleeding. It would stop soon enough. With the calf aboard his truck, Del gave a mooing call, and the cow followed him to the main herd—now a maternity ward of frolicsome calves and fertile mothers whose ebbing and flowing estrous cycles were keeping several muscular bulls hustling to keep up with their summer jobs.

Some afternoons Del even found a little free time to work on a shop project, a drive-over belly dump that he had started building almost two months before. Once finished, it would considerably speed up the unloading of his grain trucks on the farm. Factory-made units sold for five thousand dollars. He had about seven hundred dollars in used steel in this one—identical to factory-built units—but it had taken much more time to weld together than they had thought it would. Zach helped and said he was so tired of welding the steel grating and ramps that he didn't want to get near it.

Del agreed. "We're sure not going into the business of making these things. I'd hate to count the hours we've put in on it. And we're still not finished."

Later in the evening, Del hurriedly ate an early supper, talking to Caitlin while standing in the kitchen, then rushed out to take over swathing when Harlan left. Moments later, Zach phoned Caitlin. He'd finish a neighbor's fescue soon and needed to refuel the tractor at the homeplace. "Can I eat supper at your place before moving the swather to another job later tonight?" he asked.

Caitlin was used to unpredictable meal schedules. She turned to Amber, who had stopped by after work and was hoping to see Zach for a few minutes during his on-the-run supper, and said, "You might as well stay and eat with him too."

"I'll help you with supper, then," she volunteered.

"Oh, that's not necessary. I always keep something around for them to eat anyway because you never know when or if they'll show up," she said with a wave of her hand. "It may not be fancy, but they don't mind. They're not dressing for supper anyway. Sometimes they don't even sit down."

After Zach left, Amber stayed to visit with Caitlin and to help her with projects for her summer Bible school classes starting next week. Several years ago, she had volunteered to help out with the classes. "Now," she said, "it's hard to say no."

Amber took Nathaniel outside to look at Grandma Caitlin's flowers and vegetable garden again. Exploring the garden was terribly exciting to a two-year-old. By now it was a riot of green, everything sprawling and climbing beyond allotted boundaries. The garden was Caitlin's, of course, because she'd done most of the planting, but Zach had masterminded some changes earlier in the year. Using his bulldozer and a borrowed laser transit, he dozed about two feet of topsoil from the west end of the garden and moved it to the east, reversing the drainage so it sloped gently westward toward the driveway. Then, rather than using the garden tiller, he ploughed it with one of their small Farmall tractors and a two-bottom plough and worked the soil with a tractor and disk. Still not satisfied, he attached an old two-row planter to the tractor and planted six rows of sweet corn. Then he marked out the rest of the garden in rows using the planter to make dummy rows—all the same width—so he could cultivate them with the tractor too.

Zach's approach to gardening was a little heavy-handed for a thirty-five by seventy-foot plot, but it saved time. Caitlin agreed that it took efficiency to a new level. Aside from staking up some tomatoes, she'd hardly put a spade or hoe to it this spring.

By mid-June, the cherries were ripening. Caitlin had picked several gallons during the past week. One small tree on a rental farm was bent under the weight of its cherries. Caitlin didn't have time to get them, and she couldn't get anyone to pick them. It seemed a shame. She knew the birds would soon have their way with them. Somebody declined because they didn't have a cherry pitter. Someone else didn't have time. People didn't seem to value such things as much now. Had they forgotten the aroma of fresh-baked bread or the taste of garden-fresh tomatoes or tree-ripened fruit? she wondered. Even in rural America, the convenience of prepackaged products was trumping garden-fresh and homemade foods. A neighbor did bring by some huckleberries. They looked like blueberries, and Caitlin said she wouldn't have known the difference except for taste, but there wasn't enough for a pie.

There were fewer people in the community now, too, so there wasn't as much opportunity for sharing summer vegetables and produce. The ones still here were so busy farming or working off the farms that few of them grew much in their gardens. Some didn't even have gardens. The Amish were the exception. They had large families, and their gardens fed their families and provided income from roadside kiosks during the summer. They also sold fresh produce in season to grocery distributors in Kansas City and Springfield. Most mornings their black buggies could be seen parked along the highway outside of Prairie Point. They offered an ever-changing assortment of vegetables and fruit, much of it sold by their children. Their baskets of food, offered from makeshift roadside booths or from the backs of horse-drawn buggies, seemed to personify summer more than the sleek, modern tractors and swathers and high-tech combines working in fields adjacent to them.

But summer was more than Amish children selling vegetables and pies, and more than coreopsis and butterfly weed and bee-balm and black-eyed susans filling roadbanks with color, or men and machines laboring in fields, or grain elevators brimming with harvested crops. For visiting urban dwellers, summer might evoke nostalgia for a romanticized Currier and Ives rurality of the American heartland in which virtue and pastoral prosperity were static and unchanging. But summers weren't static, didn't consist of iconic images of what once might have been. Fields were now swathed and harvested with giant equipment that mitigated much of the manual labor of old. Farms were larger, people commuted to distant cities to work, and landscapes everywhere were changing. Nothing was static, not even scattered gardens, which were already in a photosynthetic arms race with summer weeds, nor the star Arcturus, which would climb high in the sky each evening only for two more months. Summer was just coming into its own now, and it was a mix of old and new juxtaposed across the countryside. Almost before anyone realized, these rich days of summer would change, sliding imperceptibly into a new season, just like the people living there.

20

June 15
Community

Del was at the MFA Co-op in Prairie Point to purchase cattle supplement, but when he backed his pickup truck up to the loading dock at the grinding facility, there wasn't anyone around to load it. He climbed the short outside stairs onto the rough concrete ramp and walked inside. "Hello? Anybody around?"

There was no answer. He guessed the employees were at the main building, maybe on a mid-morning break. Even the drive-through unloading bay at ground level was empty.

The building housed the co-op's feed grinder and sacked cattle feed. In the back of the building, rows of commercial cattle supplement were stacked high in overlapping layers. Most of it was the co-op's own brand packaged in sturdy, plain brown paper sacks with a picture of two black cows and the words "Beef Feed" in large outlined letters across the front. The big, dusty grinder was silent.

Just as well, Del thought. The thing howled like a banshee and showered everything with fine, whitish grain dust when it was running. The interior of the wood-frame building was bathed in dust, softening geometry of angled supports, rafters, and overhead beams. A fright of dust-shrouded spider webs spanned gaps and dangled from rafters like discarded yarn from a house of horrors. In the filtered light they veiled the confusion of hoppers, elevators, belts, and an enormous electric engine all dusted in gossamer whiteness like a vapid Monet.

Walking to the dimly lit back of the building, Del took a hand and brushed at the periphery of a stack of cattle supplement sacks until he located the ones he wanted. Dust obscured the lettering and flew in clouds as he swiped at the fifty-pound sacks. He'd load them himself. Making several trips, he carried a dozen sacks out to the truck. On one of the trips he looked up and noticed that on the side of one of the grinder's giant hoppers someone had written with a large black marker, "Somewhere in the world it's five o'clock," and drawn a picture of a clock below the letters with the hour hand on five. He peered up at the lettering, powdery with feed dust, and at bands of scintillating dust illuminated in stray shafts of sunlight entering cracks in the walls. He thought of the noise and the heat. It was only mid-morning and already sweltering inside the metal building.

"If I worked here, I'd probably want it to be five o'clock, too," he mumbled to himself.

As Del drove back to the co-op office, he crossed over an embankment where two sets of long defunct railroad tracks remained. He slowed for the tracks, and just beyond them met Bernie Larousse, who swerved slightly and stopped, pulling his pickup truck alongside Del's in the opposite direction. Bernie managed the fertilizer plant and ran the co-op's herbicide division, which provided crop-spraying services. Del had used their spraying service for years.

Bernie, wearing an American flag–patterned cap with a MFA Co-op logo on front, leaned out his truck window. His brawny frame and graying beard imparted an authoritative appearance that belied his easygoing manner.

"How's your fields doing?" he asked.

"We put Prowl on the soybeans when we planted," Del responded. "Pretty clean so far. I'll let you know in a week or so if I need help, but we just bought a spray rig so we should be okay. How you doing?"

"About a thousand acres behind in spraying right now. Too much rain. Can't get the sprayer in the fields."

"I imagine so. Worked ground is still real soft after the rains. But sod fields are fine. We're almost finished swathing fescue now. About ready to start picking it up."

"Some guys around here are going to lose most of their soybeans if we don't get Roundup on them soon. Can't hardly see the beans anymore in some fields. Landon's got those fields north of town with all those low places. Don't known when we're gonna get to them."

They parted with a nod and an upraised hand. It was not the business of scheduled appointments and legal contracts with guarantees or prepayment clauses, and it required no bankers or lawyers or notaries. Their casual meeting wasn't even business so much as men who depended upon each other, reaffirming what they already knew, on a street in a small town on a warm summer morning. This was business in a community where people knew each other and trusted each other, where a person's word meant something and his reputation everything.

Del wanted to see his calves sell—again. These calves, the ones he'd sold last November, were about to change hands again. He called Clint, the buyer at Mid-Continent Auction, to find out when they would be holding the next auction. Looking up lot numbers and dates of the four-day sale, Clint estimated the calves Del formerly owned would be sold to a new buyer between about ten a.m. and noon. If he hurried, he could see them sell.

Caitlin wasn't at the post office today, so she made sure the auction was showing on the television. Del wasn't nostalgic to see his calves, but their weight gain and new selling price were on his mind. He'd sold them last fall using this same nation-wide auction company that videotapes herds a few weeks before sales and sells them via satellite television. Anybody in the country with a buyer's number, a phone, and a television could watch the auction live and bid remote. The calves Del sold last November would be seven months older now, and three hundred pounds heavier, some of them more than that. Last November some of them went to a buyer that lived little more than two hours away. This time they would almost certainly go to a large confinement lot somewhere out on the plains, maybe Kansas or Oklahoma, get mixed in with thousands of others, and be taken up another three hundred pounds or so to a slaughter weight of around 1150 pounds each. They would be seasoned travelers by then, and less than a year and a half old.

In the past, Del had considered holding his calves and finishing them himself, or at least taking them up several hundred pounds more before selling, but doing so was a lot of work on top of winter chores with nurse cowherds. Watching the prices and weight gains today, he thought, might help him decide if it would be worth the effort to hold them longer. Buying was brisk. Groups of fifty to 250 calves were auctioned at the rate of about one lot a minute. Buyers watching the thirty- to forty-second video clips were simultaneously bidding live by phone. Within a day, a lot of sellers would be loading these calves onto trucks and shipping them to new locations all across the Midwest, some even farther. The trucking was easy enough, but there were costs riding along too—transportation and insurance; the labor of sorting, loading, and unloading; auctioneering; vet bills; and feed. And the costs started all over again each time the cattle changed hands. Today's prices held in a narrow range, about ten to fifteen cents less than what Del had received last November. He made some back-of-the-envelope calculations—feed, veterinary costs, weight gain, selling prices, then frowned at the figures. The guy that bought his cattle last fall could have lost money on them. It was a real-life lesson in economics and the capriciousness of commodity prices. Maybe he'd reconsider finishing out yearling calves.

After his vicarious brush with the economics of feeding fat calves, Del left the house, figuring he had time to take mineral supplement to his most distant herd of cattle before dinner. The cattle were in a pasture about eleven miles away. While he was there, he could check on a cow and her new calf, maybe confine them inside the small corral in the pasture if there was a problem. The pasture was bordered by a quiet, unpaved road, and the corral was inside the pasture and some distance from the old wire gate entrance. It wasn't a fancy corral, just woven wire supported by some crooked Osage orange posts. A couple trees outside of the corral provided a bit of shade for loitering cattle on hot summer days. Beyond it was a pond with water so Del didn't need to check the herd often. Most years the corral only got used in the spring and fall to sort and load cows and calves. Otherwise he seldom paid much attention to it. At least that was the case until he found marijuana growing in the corral a few years ago.

Since the discovery of that clandestine crop, he'd made a point to check the pasture and corral more often. This morning he was reminded of it when he spotted the cow and calf standing in the shade of a tree outside the corral. When he first saw the marijuana, he could hardly believe his eyes. He'd heard it grew wild in places up north along the Missouri River, but not in this area. He had called the sheriff, thinking he would send someone out to cut it down and remove it right away, but as far as he knew, neither the sheriff nor any law enforcement official ever showed up. Maybe there was a stakeout, maybe not. He didn't know. The plants weren't mature when he found them, but the summer was young then, and the marijuana grew tall and robust in the warm, manure-rich soil. Del grudgingly admitted to himself that the corral and its rich soil had been a clever gambit, but he'd been uneasy about it being there. Aside from legal issues, he felt violated, angry that someone had used his land without his knowledge or permission. He knew he ought to chop it down, or mulch it with a tractor and brush hog, but it was a long drive—almost an hour on a tractor. Besides, a brush hog, even his smallest one, couldn't be used easily inside the cramped confines of the corral, and he figured he'd still have to chop a lot of it by hand. That, he thought, ought to be a job for the sheriff's department.

Later in the summer, the crop vanished. A good harvest for someone, Del reckoned. He felt relieved when it was gone. Since then he'd looked in on that corral more often, but the surreptitious gardener who plied his illicit trade that summer never returned. Perhaps he took his farming talents, however modest, elsewhere. In a community where neighbors know everyone and everyone's business, an unfamiliar automobile traveling a country road would eventually attract attention and provoke speculation. Yet the clandestine activities of this unknown gardener were apparently never noticed. Perhaps he worked at night. There were only two houses along that road, and neither was close. Del knew the occupants. They hadn't seen anything suspicious. Freed from the societal constraints that accompany responsible behavior and land stewardship, the activities of this rogue gardener stood in stark contrast to those of the community and to rural ethics generally. But, as Del could see, even in rural areas, a few cracks were appearing in those ethics.

21

June 25
Breakdown

"If you use a combine, sooner or later you're gonna have to repair it," Del once told Zach with grim humor. Del's observation was true for almost all farm machinery, which was routinely stressed by heavy loads, rough terrain, and dirty, dusty conditions. A combine was one of the most complex of all farm implements, capable of harvesting crops ranging in size from tiny clover seeds to foot-long ears of corn. It could cut, thresh, clean, and separate grain, and discard the rest of the plant material, all in a smoothly integrated process that took only seconds. From front to back a combine was a whirring and shaking mass of augers, chains, bearings, belts, pulleys, elevators, and sieves. Almost everything on a combine moved, and if it moved it was likely to eventually wear out. No one knew this better than Del, whose mechanical skills kept three older combines and a huge fleet of used trucks, tractors, and farm machinery running season after season.

Yesterday it had rained. Northwest of the Montgomery homeplace the rain was light, barely a quarter of an inch, but a few miles to the southeast it was nearly an inch. Del had wanted to finish harvesting fescue in these east fields before bringing the two combines home. Now he couldn't. The soggy windrows of fescue were beaten down into the stubble. They'd need two days of sun to dry, maybe more. He'd have to move the combines back to the homeplace and restart harvesting tomorrow in drier fields. This morning a few gray white clouds sailed across mostly blue skies and moved off to the east. There was a light southwest breeze, a good omen.

In one sense the rain provided a needed break. After more than a week running the combines hard, they were in need of repairs. A water pump on the 815's engine had been leaking for several days. Harlan watched the engine temperature and kept a five-gallon jug of water in the combine cab and another by the gate at the field for refills. Del had a replacement pump ready, but as long as they could keep harvesting, he postponed the repair. Replacing the pump wasn't a big job, but each night they took loaded trucks to the seed company in Grand Forks, and the lines of trucks waiting to unload were usually long. Sometimes they didn't get back until eleven, even midnight, bone tired after workdays that stretched to eighteen or nineteen hours, and they didn't feel like climbing around on top of a combine to make a repair in the middle of the night. So they pressed on knowing that eventually there'd be a rainy day.

This morning, after replacing the water pump, Zach and Harlan discovered a bearing burned out in the beater, a big rotating fan that sits behind the threshing cylinder inside the combine. The beater looked like a small version of an old-fashioned paddlewheel on a riverboat. It directed the flow of straw rearward from the threshing cylinder to the straw walkers. The dry bearing squealed in protest when they started the machine, and in a few minutes it began to get hot, a danger around dry straw and chaff. By late morning they'd replaced the bearing, removed some links in two slack chains, greased and fueled both machines, and cleaned the air filters.

Del, meanwhile, made a trip to Bellamy to talk with a judge. He had promised Harlan he'd help him get some legal advice on his marital problems. Harlan was on the outs with his wife again. He swore this was the last time, but there'd been other "last times."

"Harlan, you ought to just get a divorce," Del told him. "All you two ever do anymore is fight. She thinks of things to make you mad. Then you turn around and think of something to make her mad."

"I know," Harlan acknowledged, fidgeting with his cap and clenching his fists. "I don't understand why she don't just leave me alone. I can't put up with this shit no more! I just want to get my stuff out of the house. But I gotta make sure there ain't nothin' that prevents me from going there when she's gone."

Harlan was so upset he threw his cap on the ground and stomped off into the field behind the combines, then looked away, staring far off, face a torment.

When Del got back, he called Harlan over. "The judge says there's nothing that prevents you from getting your personal effects out of your house. But you can't do it when she's there."

"That's what I thought," Harlan blurted out.

"Listen, you need to make a plan. Wait until she's at work. Not this afternoon because you don't have enough time. But maybe tomorrow morning. Make one trip over there and get all your stuff, and I mean all of it. Then get out of there. Don't go back claiming you forgot something. You hear me?"

Harlan heard the words, but Del wasn't sure he'd do it. Later he said to Zach, "I hope he don't mess this up. It'll be a shame if he does 'cause he's a good worker when he's around. He's got a lot of talent for mechanics and things."

Del walked around the two combines, his eyes roaming over the familiar shapes of belts and pulleys and hydraulic lines. He spotted another problem. The seams on the variable speed thrashing cylinder belt on one of the combines were beginning to separate in a couple of places. It was an important belt, providing power to the thrashing cylinder and controlling its speed.

"Looks worse than it is," Del said. "The outer part of the belt is the strongest part. It will probably hold for several days, maybe longer. Might last 'til the end of harvest."

Then Zach noticed a leak in a hydraulic line on the 915 combine. The leak also didn't look bad, just a small greasy area where chaff and dust had collected. Del examined it. "Might be an o-ring seal leaking. We got a box of them in the truck, don't we?"

Zach looked. "Nothing here. We must've left them in the shop." They decided to "road" the two combines back to the homeplace, about five miles away.

"Fescue will be drier there anyway 'cause those fields didn't get much rain. Maybe we can restart there tomorrow. Give us time to eat and fix this after we get home," Del said with a shrug.

Zach started back to the homeplace, driving the lead combine. On a long downhill slope on the county blacktop road, he saw a car in the rear-view mirror of the combine. It was approaching fast. He started to slow the combine and ease it onto the sloping edge of the road to let the car pass. But something was wrong. The combine wasn't responding. Combines were unwieldy, top-heavy machines, and they steered from the rear. Even worse, the bulk of a combine's weight was up front, making it especially unstable traveling downhill where they were prone to tipping forward. For that reason, operators usually used the belt-driven, variable-speed transmission to control the machine's speed, slowing it gradually rather than using the brakes. Zach pulled back hard on the transmission lever. Nothing!

Now he was actually gaining speed, making the combine even more difficult to control. The only way to slow it was with the individual front wheel brakes, which were designed mainly to assist in turning in fields, not stopping at road speeds. Pressing one harder than another could cause one of the big drive wheels to lock, causing the combine to swerve violently, even flip. Pressing both too hard could lock both front wheels and cause the back end to rear up and the combine to tip over on the sloping roadside. More than one farmer had totaled his combine that way, rolling it into a ball of scrap metal in a ditch.

Zach had to get the combine under control quickly because now the driver of the car behind, not realizing the danger, was starting to pass. If he lost control of the combine it could lurch to the side and hit the car. He had to use the brakes to slow the runaway combine. He set his heels hard against the floor and touched the left and right brake pedals simultaneously, very lightly, trying to apply the same pressure to both. The brakes shuddered. He pressed harder and the combine began to vibrate, then escalated into a noisy, violent shaking, but the brakes didn't lock. The huge machine began to slow.

The driver of the car, impatient to get past the big, slow-moving combine that was blocking much of the road, charged ahead, running partially off onto the grassy left shoulder of the road to pass. Zach was so intent on keeping the combine under control that he scarcely noticed that the driver of the car, temporarily losing control, went careening off the road and whipping through the waist-high fescue grass on the opposite roadbank, sending a shower of seeds into the air. The car skidded sideways in the slippery grass, then abruptly popped back up onto the blacktop as the driver struggled to regain control. Without so much as a rearward glance, the driver roared off down the road, a grill full of fescue stems and a storm of seeds blowing off the hood of his car.

"Crazy fool," Zach mumbled through clenched teeth as he battled for control of the combine.

Near the bottom of the slope, the combine rolled to a halt, and Zach eased it onto the sloping roadbank. For a moment he sat, unmoving. A flood of tension drained from his body. Only then did he realize how hard he'd been gripping the steering wheel. Hands numb, face sweaty, he slowly reached down for his cellphone.

"Hey, Dad," he said in a voice more calm than he felt. "You need to bring a truck and a chain. I'm on the side of the road about a mile east of the homeplace. I think one of the hydraulic lines must've busted. Transmission wouldn't respond, and I thought I was going to ride this thing into the ditch. Crazy driver passed me right in the middle of it all. Lucky I'm still right side up."

In the combine's rearview mirror, Zach now saw Harlan approaching in the second combine, the top of his cab just coming into view over the crest of the hill behind. He quickly called Harlan and told him not to stop but to continue on to the homeplace.

"Dad's on the way," he said.

When Harlan slowed and eased his combine around Zach's disabled machine, Zach saw a thin stream of diesel fuel spewing out of a tiny hole in the fuel tank of Harlan's combine.

"What next?" he muttered, shaking his head.

He called Harlan again. Harlan said he'd already seen the leak and was going to try to plug it as soon as he got to the homeplace. "I just topped off that tank this morning. Almost thirty-five gallons. Don't want it all to leak out."

The chain Del brought wasn't long enough to reach forward under the combine header and pickup attachment, so he couldn't tow the combine forward and pull it home. Instead, he towed it backward up the hill until they found a level place to park it. Once the combine was safely off the road, Zach and Del looked at the hydraulic line again. The problem wasn't an o-ring leak after all. The steel line itself had a leak, a hairline crack that had opened under pressure, causing the fluid to leak out.

At the homeplace Harlan set a bucket on the ground to catch the leaking fuel from his combine and then held a finger against the hole to temporarily stop the leak. When Del arrived, he told Zach to get a drill and a small screw with a rubber washer. A narrow dripline of oily red diesel slowly made its way down Harlan's forearm. "And bring a rag," Del hollered to Zach, who was already in the shop. "I don't want this stuff squirting all over us when we put the screw in the hole."

"Can't you drain the tank with the valve at the bottom?" Zach asked.

"It's stuck," Harlan replied. "I already tried that. That's why I'm here like a little Dutch boy with his finger in a hole in the dam."

"Harlan," Zach said, laughing, "ain't no way anybody's gonna mistake you for a little Dutch boy."

"Here, hold the rag up around the hole. I'm going to drill it out," Del said, taking the drill.

Harlan released his finger from the hole, and diesel squired onto the rag and the drill. It splashed across Del's shirt and jeans and ran down his tanned forearms, but the hole took only a second to drill. Zach then immediately jammed a screw with the rubber washer in the hole and screwed it tight.

"That'll hold it until we can get a bolt in it and make a better seal." Del wasn't surprised at the leak. It had happened before on this combine, and on the bigger 915. He figured it was a manufacturing flaw. The baffles welded to the inside of the fuel tank weren't strong enough for the weight of the fuel.

In the kitchen, relaxing for a few minutes over Caitlin's pork chops and fried zucchini, the men passed plates of food and rattled glasses of ice tea and discussed mounting mechanical casualties. They would try to find a hydraulic line and a variable-speed thrashing cylinder belt on one of the hundreds of old junked combines at Crockett's Machinery Sales this afternoon. The leaking fuel tank could await a more permanent fix later.

At Crockett's, they searched up and down rows of junked combines, looking for the right model. There were hundreds of salvage combines, a graveyard of oddly shaped, cannibalized skeletons of steel. Most were old and worn out. A few had caught fire and burned in fields, probably when a bearing or some other moving part overheated and caught tinder-dry straw on fire. It wasn't uncommon. Still, most of the machines had salvageable parts. Some were heavily cannibalized, bearings and pulleys cut out with torches, engines and wheels removed, cylinder bars stripped, insides gutted, the combines themselves propped on blocks, some tipping against others or listing dangerously. Eventually they found an almost-new variable-speed pulley belt and lugged their toolbox over to the combine and removed the belt. But they couldn't find a steel hydraulic line the right size or one that hadn't been damaged. Finally, they realized they'd have to fix the one they had.

Del called Hayden's welding shop in Belle River. Could he braze a hydraulic line this evening if they brought it by?

"When?"

"About an hour."

Hayden said he was going fishing with his grandson. Still had to get his boat and fishing gear ready. If they weren't too late, he'd do it.

Back at their combine, still parked along the roadside, Del and Zach worked quickly to loosen clamps and couplers, scuffed knuckles on greasy bolts, grimaced and pushed, but to no avail. The line was stuck. They'd have to unbolt and remove the fuel tank first to free the hydraulic line next to it to get at the damaged one. "If them dang engineers that designed these things had to fix them, they wouldn't do dumb things like this," Del complained. "Have to disassemble half the combine just to remove one little hydraulic line."

Glancing at their watches, they pressed ahead, eventually removing interfering parts to get to the hydraulic line without damaging it further. They reached Hayden's welding shop just as he was hitching up a small fishing boat to his truck.

Hayden had been a professional welder most of his life, working out west. Several years ago he had returned home to open a shop in Belle River. His skills were legendary. Farmers and construction outfits brought him all their difficult jobs. A large man with a broad smile and laughing eyes, he wore a red and white brocade-patterned baseball cap with the bill turned backwards so his welding helmet and goggles would slip on and off easily. A large denim shirt hung like an apron halfway to his knees. Jeans and heavy brown steel-toed boots protected his legs and feet from hot slag and errant flecks of sharp metal. Hayden was a master of welding mediums—migs, tigs, plasma, brazing, and silver solder—and he was also a preacher. Someone once joked that there was fire and brimstone in his shop and even more in his belly. There was no doubting the sincerity of his calling. When someone needed his preaching services, day or night, he closed up shop. If someone had a welding job, it paid to call ahead. Del had found that out the hard way.

The shop was an anarchy of steel—heavy angled plates, steel bars, steel cylinders, steel scrap, some of it neatly stacked, most of it geometric bedlam. Hundreds of feet of dusty hoses from oxygen and acetylene tanks snaked back and forth in crazy eights across the grimy floor and eventually linked pressurized tanks with torch nozzles. More tanks and hoses and steel plating were propped against the back wall.

While Hayden prepared to braze the crack and the coupling on the end, Del called a farm supply shop in Ashton. It was late. Everybody else was closed.

"Yes, we do have hydraulic fluid," a voice on the other end answered.

"Can you set five gallons of fluid outside the back door of your shop? I'll pick it up later this evening."

"We've only got four gallons."

"Set out what you got. It's for Del Montgomery. Uh huh. Montgomery. Yep, with an M. That's right. Del Montgomery. I'll stop by and pay for it next week."

Hayden worked with confidence, degreasing the metal hydraulic line, adding flux, heating the line so the filler metal would flow over the small crack and seal it just right, and finally reheating the line in hot water to scald off the flux.

Zach watched closely. "He's good," Zach said, his voice showing admiration for skills that exceeded even his own considerable expertise as a welder. "There's almost nobody around here as good as he is. I wish I could braze like he does."

Hayden did his work over a heavy steel table at one side of the shop. As he continued, now attaching the coupler, his torch illuminated the dimly lit shop in flickering blue light that cast dark shadows on the wall. His torch hissed, and a few flecks of hot metal and sputtering flux fell onto his worktable. Minutes later, he pronounced the line ready for service again.

Hayden's office was in an adjacent room. It was separated from his shop by a dusty, waist-high wooden partition. The office was almost as large as the shop, and visually the transition from welding shop to office was minimal. Boxes and racks of metal pipes, couplings, and electrical fittings filled the room. There was a small walk space in front of two desks, the desks themselves mostly buried beneath even more dusty boxes, welding manuals, and cylinders filled with a bewildering array of welding rods. Two rusty, four-drawer file cabinets with several large cardboard boxes precariously balanced on top stood beside one of the desks. Hayden used the only available space on the desktop surface to prepare the bill by hand. Del handed him a yellow bank check, and Hayden laid it on top of a stack of papers and manuals on the desk. Almost as an afterthought, he laid two loose welding rods atop the check. Del wondered how he ever kept track of anything in there.

Del and Zach were back along the roadside by seven, installing the now repaired hydraulic line on the combine and reseating the fuel tank, each part replaced in approximately reverse order from which it had been removed. Zach wiped his forehead with a grease-stained hand and looked at Del. "You know, I'm more tired today than if we'd been running all day in the fields. Chasing down parts and repairing things all over the county is worse than harvesting."

Del nodded in agreement.

And they weren't finished. He had to drive to Ashton and pick up the hydraulic fluid. The four gallons would be enough to get started. He'd get more somewhere else tomorrow. Zach could install the variable-speed cylinder belt later that night. They figured it would be ten or eleven before they finished.

"And we haven't looked over the 1680 combine yet. It's been parked in the shop since the rain," Zach added.

The sky was mostly clear, but rumbles of clouds were building in the west, dark and dimpled underneath in the gathering dusk. Below the clouds, on the western horizon, there was a seam of red, like a hot weld fusing violet sky to stygian earth, the last hint of the spent day. They hoped the rain would hold off for three more days. If it did, and there were no more breakdowns, the last three hundred acres of fescue harvest would be history. They knew that when you lived by the land you lived by hope.

22

June 30
Summer Harvest

By the end of June, the sun reaches its northernmost passage and begins slipping southward, the intensity of its rays fractionally weaker each day. It wasn't enough to notice, not yet, anyway. There was no change in the hot, steam-in-your-face humidity and soaring temperatures of midsummer. Days would remain hot for some time, even increasing as the vast continental interior continued to gain heat, but over time the delicate balance of the heat equation would shift, sliding inexorably toward winter. It seemed too early to think about winter now, but the passing of the summer solstice defined, in many ways, the pinnacle of the growing season. If the days leading up to the solstice encompassed an exuberance of energy, of life bursting forth from Mother Earth in a vast orgasmic spasm, then the days following seemed to coast, skating along on pent-up energy that, for a time, propelled summer's excesses to greater heights. But soon, neighbors would mow their lawns and roadside less often, the air would feel drier, the symphony of morning birdsong would fall silent, and flecks of brown and gold would bring summer fatigue to pastures. Farmers, too, would soon know if their harvests were good and if the labor of their carefully planted crops would meet hopes and expectations.

Zach, looking at a burgeoning cornfield, said to a neighbor, "Corn's really flexing its ears now." It was an affirmation of the favorable growing season so far this summer because corn was sensitive to drought. This year there wasn't a hint of curling or burning on the long, narrow leaves. Not yet, anyway. The corn rustling softly in the morning breeze stood shoulder high in early-planted fields, and stalks showed silky tassels. Ears were already long with thick husks and large kernels. Even so, rain in the next week or two could add another ten bushels an acre to the yield. That's what Del was hoping.

June was a busy month for midwestern farmers, one of the busiest of the year. Fields teamed with activity—trucks, tractors, swathers, combines, mowers, hay balers. Farmers looked like farmers—baling fescue hay, moving bales to storage, threshing grain in fields, trucking harvests to elevators late into the night. They grumbled, and some cussed when things didn't go right, and they were often fatigued. And their jeans and overalls were often dirty, too, but you could see in their stride, in the way they carried themselves, that they were proud of their accomplishments and glad to be getting on with their work. Except for a couple of Saturday night tractor pulls, Del and Zach hadn't had a free day all month, not since spring planting.

A few days before, Buddy Dean's combine caught fire, and then it burned up the wheat crop he had been harvesting. News of the fire traveled quickly. Someone wondered if he had insurance on the combine or the wheat. Buddy Dean burned his arm trying to put out the flames engulfing his combine, but he had to retreat when the fire spread to his field and he lost everything. Prairie Point's volunteer fire department arrived promptly, but they could do little except set backfires to contain the burn. A volunteer firefighter opened a small access door on the combine, and the rush of oxygen explosively reignited the fire inside the machine, and it singed his eyebrows and eyelashes. Del said they had a combine fire two years ago, but they had enough fire extinguishers in the combines and trucks to put it out before it spread. The fire was caused by a bearing that overheated and froze up.

"Probably a bearing on Buddy Dean's combine," Zach said. "We were lucky. All it takes is one bearing overheating that you don't notice."

"More than just bearings cause problems. Straw wrapping around a shaft can get hot and start a fire too," Del added. "Like a driveshaft on a car or truck out here in a field with lots of straw and ever'thing."

Inside their farm shop, Del and Zach finished repairs on their combines and discussed how to move them to their last fescue fields, a for-hire harvesting job several miles southeast of Prairie Point. They had hoped to be done by now, but rains had forced them to move to drier fields. Now they had to move the combines back again to finish.

It was almost as much work to move the combines to jobs as it was to do the harvesting. Modern combines were awkward and slow moving, and especially dangerous on highways because of their width. Zach had experienced that firsthand a few days prior. Wide headers usually had to be removed and transported separately on highways. The small headers with pick-up attachments used in fescue harvests, however, made moving or transporting the combines easier. But there still could be problems even on county roads where bridges weren't wide enough to let a combine header pass, or the bridge banisters were so high they couldn't raise the headers above them. Earlier, by the farm shop, Del and Zach discussed a longer route that minimized driving on paved highways and avoided the county bridge problem.

"How'd you get the swather into the second field down there?" Del asked Zach.

"I went over that little hill between the fields. There's a track, but you have to watch for rocks."

"Can I get a truck in the field from the main road?" Del asked.

"There's no gate along the main road."

"Can you call the owner and ask what's the best way to get a truck in there?"

"He's gone. But past the farmhouse there's a gate that should be big enough for a truck," Zach said. "Otherwise, you'll have to backtrack around and go over that hill."

"What about the combines?"

"Gate's too narrow. We'll have to take them over that rocky hill," Zach said, raising his voice slightly to make himself heard above a sudden outburst of noise from house sparrows still trying to nest in the overhead track of the large shop doors.

Outside the shop, Zach drove the 815 combine up beside the three large fuel storage tanks, opened the combine's diesel tank, and switched on the electric fuel pump. With the tank filling, he went around to the back of the combine, wiggling his way up inside the dusty overhanging rear end of the machine. Stretched out face down inside the combine on top of the straw walkers, with only the toes of his heavy boots visible, he reached forward in the dim light and felt for the hidden grease fittings on the straw walkers. Older combines, like this one, had dozens of bearings and movable parts that required regular greasing, but many were inconvenient to reach. The newer 1680 rotary was a big improvement. Only four fittings on the entire combine required daily greasing; others could go fifty hours without attention.

Shortly after nine in the morning, Del and Zach left the homeplace, each steering a combine down the county blacktop road and into the morning sun. Even with the small headers, the combines spanned three-fourths of the road width, so the men were careful to keep the right-hand wheels as close to the shoulder as possible. They passed several farmhouses and two gravel roads leading off to the left. At a T-intersection, where a house sat at an angle to the junction, they turned south toward Prairie Point, continued straight through to the far side of town, and then turned east again. At the southeast corner of town, they exited to the south on a gravel road for a mile and a half, then east another mile on gravel, emerged onto a paved highway, and drove on the shoulder of the highway for three more miles before turning left and crossing the four-lane federal highway. On the opposite side, they followed a driveway that led past a farmhouse and down a hill to the first of two fields of swathed fescue—the last of the season. The combines, moving at ten to twelve miles per hour for most of the route, took over an hour to make the journey.

At this time of the year, most of the roadsides hadn't been mowed, and wildflowers bloomed in abundance, stippling the route with flecks and splashes of color. Scrubbed fresh from the previous night's rain, dewy raindrops on flowers and grasses sparkled like diamonds on bits of white, yellow, lavender, and orange in the morning sun. A few spike-like bouquets of white wild indigo, a tall, open plant, grew on dry roadbanks. Gangly compass plants, a sunflower with deeply cut, rough-to-the-touch leaves, rosin-filled stems, and cheery yellow flowers, rose almost as high as the back of a horse in a couple of nearby prairies. True as a compass, its flowers were pointed toward the morning sun and would track the sun's westward journey through the day. Less statuesque but no less elegant, wild bergamot bloomed in mauve patches that contrasted with the intense orange of butterfly weed, whose roots once were used in teas and tinctures to treat lung inflammation. And everywhere along roadbanks, a colorful array of lesser faces, of fleabanes, vervains, coneflowers, nettles, spiderworts, parsleys, pokes, bedstraws, clovers, chicories, and St. John's worts, some tiny, some bold, pulsed with the energy of summer on these narrow roadside stages.

JUNE 30. SUMMER HARVEST

It wasn't a big job. The Montgomerys could have harvested the sixty acres of fescue in a few hours, but the windrows were wet from the previous night's rain and would need another day of sun to dry. Del and Zach left the combines by the side of the first field and rode home with Harlan, who'd come to get them. They'd return the next day.

Del drove his twin-axle truck and gravity-flow trailer, both loaded with fescue seed, to the co-op in Prairie Point later that morning. With the weather clearing, he'd need an empty truck to finish the harvest. Driving the truck onto the long scales at the co-op, he set the air brake and waited a minute until the girl inside the office signaled through a large window that she had recorded the weight of the truck. He then circled behind the metal Quonset building and elevator with its convergence of ten-story-high augers that funneled down like stilts to several large grain bins. Doubling back, he stopped in front of a flat, concrete unloading area where a two-wheeled conveyer-belt elevator extended up, like a pregnant stick insect, over a waiting tractor-trailer. Two more empty tractor-trailers were parked nearby. By the end of the day, all three would be filled and headed to a seed company twenty miles away in Grand Forks. Fescue seed was light and fluffy and plugged Archimedes' screw augers so it had to be loaded using a slower belt elevator. Del's truck unloaded the seed much faster than the little elevator could carry it away. Soon a large brown pile of spilled seed began to accumulate on the concrete slab behind the truck. When the truck was unloaded, he pulled forward and repeated the unloading process with the gravity-flow trailer towed behind. It unloaded even faster, and another large pile of spilled seed quickly accumulated on the concrete.

Three teen-age boys in ball caps, jeans, and t-shirts with the sleeves torn out at the shoulders struggled mightily to scoop up the ever-expanding pile of spilled seed. With slender brown arms and pencil-thin legs, they scooped furiously, unmindful of the envelope of itchy dust surrounding them as they strove to contain the spilled seed. Hired by the seed company for the month-long fescue season, they had seen little but fescue seed and truck tailgates for several weeks. There were no days off. Banging hard on the side of the gravity wagon, one of the boys could tell by the dull thud that seed remained lodged inside the hopper. He banged again, harder, but the clumped seed stuck to the sloping sides of the bed, not sliding freely like heavier corn or grain. Frustrated, he climbed to the top of the gravity-flow trailer, swung a leg over the side to hold himself in position, and shoved the last of the seed into the bottom of the hopper with his shovel.

Del weighed the truck and trailer again, this time empty, so the co-op could calculate the weight of the seed. When he left, another truck was blocking the road, and he had to circle the co-op again. As he ran the transmission up through a few gears and swung the truck onto a side street leading back toward Main Street, he passed an old building on the left, a long-defunct lumberyard. The front of the building had been partly repainted with fresh white paint. Large block-shaped, black lettering on its side proclaimed it the world headquarters of a cheeky little publication that ran advertisements, announced local sales, dispensed doses of wisdom and wit, and, if nothing else, showed considerable grit in its ability to survive the economic ups and downs of small towns. It was a one-family operation. The man and his wife were owners, editors, and publishers. They had been at it for more than two decades.

Main Street was barely over three blocks long and dead-ended at the sprawling co-op with its supply store, elevators, grain storage, fertilizer, and feed-grinding facilities, so Del passed by the town's main street each time he stopped at the co-op. During harvests, that was sometimes several times a day. Main Street was dying. It was, like those in many small midwestern towns, a dowdy mix of two-story, red-brick buildings dating back a hundred years or more. A few newer single-story buildings with metal fronts were also in the grip of hard times. When the town was incorporated in the late 1800s, twenty businesses applied for licenses, sidewalks were wooden, Main Street was dirt—or mud when it rained, coal oil lamps lighted the streets each evening, and daily passenger and freight train service was heralded with much anticipation and large gatherings of residents. Almost everything arrived by rail, and lively fairs and celebrations on Main Street attracted hundreds of people.

Today sidewalks were concrete, streets asphalt, the lights electric, and the railroad, along with most businesses, ghosts of decades past. Some businesses had moved out alongside the new highway east of town. Only a grocery, restaurant, furniture store, and post office remained along Main Street, and the grocery and furniture store were about to close. The town would soon be a food desert for those unable to drive elsewhere.

Del had seen many businesses flourish, then die. When he was in high school, there was a lumberyard, bank, tavern, grocery, two auto repair shops, tire shop, electrical shop, feed store, blacksmith and welder, even a movie theater. All were on or adjacent to Main Street. For years there was even a resident osteopathic doctor. People came to town on Saturday nights to visit and to do business. His father ran the grocery. It had been one of the survivors. After his dad died, Del's older brother ran it. Now a nephew ran it and would preside over its final days. Growing up here, Del had delivered groceries to people all over town. He knew the names of almost everybody in town, all five hundred of them. Since then, the population had barely changed, revised upward only slightly. He still knew the names of many of those former inhabitants and where they once lived, but now most of them were gone. Only a couple of people still lived in the same house they did when he delivered groceries.

Once Prairie Point was a thriving little economic center, but nowadays people were more mobile. They no longer needed small towns. Prairie Point became top-heavy with elderly residents—retired farmers, widowers, older folks from surrounding rural areas who moved to town when they were no longer mobile. Today there was a rest home for the elderly, and the town struggled with growing numbers of welfare recipients whose properties were often unkempt. Many of them had come from elsewhere and lacked the pride of a shared community history. Lately the county sheriff and his deputies had made meth lab busts in the town. Del knew that drugs had crept into small towns. Prairie Point wasn't immune.

There were still plenty of people in town with civic pride. The history of most small towns included rough patches, but the town's economic pillars—as a business and agriculture hub—were badly eroded. Some towns gained in prosperity, but often at the expense of others. Rural areas had been losing populations for decades. Once Prairie Point served hundreds of farm families in the area, but now barely a dozen survived entirely by farming, and fewer than half of these practiced the kind of mixed row-crop, grass seed, and cattle farming that Del managed. Some rural residents worked at off-farm jobs, commuting long distances to big cities. There were a few semi-retired farmers, or those retired from city jobs. They often kept a few head of cattle on small farms, but their activities were as much hobby as necessity. Farmers whose income came entirely from agriculture operated on a larger and more efficient scale—far larger than their fathers or grandfathers could have even dreamed. Their business and economic ties stretched beyond the local community, beyond the county, even beyond the state. Like Del, they watched profit margins decline and were forced to expand and streamline their farming business to survive. Kids weren't staying around, either, so there might be even fewer family farms in the future.

A block west of Main Street, Del turned right at a street corner, shoved the heavy truck transmission up through several gears as the engine alternately accelerated and coasted, and headed home. The town's past was behind him. He was already thinking about the afternoon's work.

After his noon meal, Del climbed up the metal ladder that extended out over the big left front tire of the 1680 rotary combine. Scooting onto the wide padded seat inside its glass-enclosed cab, he revved the engine and began the one-hour drive west on the county blacktop road to his most distant property, and to his last unharvested field of fescue. With the other two combines now parked at a neighbor's field far to the east and awaiting drier conditions, he'd harvest this field with his remaining combine. Zach and Harlan followed in trucks and left the big International cab-over semi beside the field. The land to the west of the homeplace flattened and opened up some. From the big front window of the combine cab, he had a fine view of the countryside spread out in a panorama of well-ordered fields, but his focus was on keeping the giant combine out of harm's way of traffic. Like guiding an elephant up a sidewalk, he reckoned, but there was little traffic today.

Leaving the highway, Del headed south a short distance. Passing through a wide, double gate painted red, he drove the combine across an open pasture, past scattered Black Angus cows and calves, and toward the far side of the pasture where a forty-acre field of fescue was enclosed by a barbed-wire fence. Toward the back of the pasture, he passed an old pond dug into a gentle slope, and on this warm summer afternoon a half-dozen black cows stood belly deep in brown, addled water. He smiled as he passed them. If the water hadn't been so muddy, he might have enjoyed a cooling swim himself.

Inside the fescue field, he lowered the combine's header and pick-up attachment and increased the engine's RPMs while simultaneously checking cylinder threshing RPMs and listening to the machine's familiar rumbling sounds as it reached harvest speed. He shoved the variable-speed transmission lever forward, and the big machine eased ahead. Almost immediately, it began devouring a windrow of swathed fescue at a pace faster than a human could walk. The windrow wrapped around and around the L-shaped field, gradually spiraling inward like a giant labyrinth puzzle. Only near the center did the windrow finally end. There, where an odd-shaped block of fescue remained, the last windrows were swathed in a series of progressively shorter point rows until the field was finished. The combine rumbled steadily, occasionally making groaning sounds when a heavy shock of grain hit the threshing cylinder. Once, a barely fledged meadowlark with a stubby tail rose in feeble flight just ahead of the spinning pick-up attachment at the front of the combine. A short distance ahead, it dropped into the stubble, then as quickly rose a second time, fluttering a few more feet before tumbling headlong into the grass and stubble as the giant machine passed over it, abruptly blocking out the light. Dust settled on the stubble, and then the machine was gone, its giant wheels passing inches from the crouched and uncomprehending young bird.

During the next few hours, Del stopped several times, pulling the combine over to the side of the field to unload the hopper of seed into the truck. It was perfect harvest weather. White clouds were stacked like fleece across a royal sky. His white truck, brimming with brown fescue grass seed, was parked at the edge of a green pasture. A second empty truck was parked nearby. Glossy swallows with needle tails and pale bellies skimmed tirelessly back and forth across the field as the big red combine, trailed by a dingy cloud of dust, wound its way round and round the field.

By five in the afternoon, Del had finished the field and was back at a large tin-covered machinery storage building a mile from the homeplace. With the combine engine still idling, he called Zach on his cellphone. "I'm at the Somerset place," he said, speaking loudly over the engine noise. "Come over and help me switch headers. Should be time for you to stump some fescue in those two pastures on the Farley place."

By now any unswathed and uncut fescue grass seed was mature and dry. It was also likely that a lot of the fescue seed had already shattered to the ground—knocked loose by rain and wind or cattle. Nevertheless, if they had time, Zach and Del often tried to retrieve what remained. They called it stumping, but, by any name, it amounted to cutting the seed heads directly with a sickle in the combine header's cutting bar. Whatever they could retrieve would be extra spending money—a few hundred, perhaps a thousand pounds of seed. Last year they pocketed almost fifteen hundred dollars from stumping fescue seed still standing in cattle pastures. Some years there was almost none.

Del and Zach quickly disconnected hydraulic hoses and unhooked the small, stripped-down header and pick-up attachment, secured it on blocks, and backed the combine away. The big machine, now bereft of its header and pick-up, looked oddly unbalanced, an emperor suddenly without its clothes. In its place they attached a much larger, twenty-five-foot table header with sickle and cutting bar and paddle-like batting reel. Del coupled the various hydraulic hoses that raised and lowered the header and the batting reel. He accidentally reversed two of the hoses, and when they tested the header, it moved up when it should have gone down and vice versa.

"Hey, you got it backwards," Zach chided, grinning.

"No way of telling which hose goes where 'til you try it," Del countered, a hint of annoyance in his voice.

"Sure there is, just look at the picture."

"Ain't no picture."

"Yeah, there is. Right there on the side of the header."

Del rubbed his fingers across the sides of the header, brushing away the dust. Sure enough. There was a diagram, or had been. Scratched and faded over the years, it was barely discernible now.

"Okay. I got the hoses switched," Del said. "Try it now."

"Good. But what's that banging noise under the snout?"

Del yelled to Zach to turn off the engine. "Don't know," he frowned, and then peered under the snout. "Here's the problem. This spring's in the way."

"More than that, I think," Zach added as he came down for a look. "Snout's been sprung to the side. See. It's crooked. We need to bend it back."

For ten minutes they pried and hammered on the snout, a conical piece of metal that covered several moving parts at the outside edge of the header and provided a smooth boundary for the cutting sickle—stems of grain swept to the inside of the snout would be cut, stems sliding to the outside would be left standing in a sharp, even boundary. Eventually everything worked to their satisfaction.

Zach guided the combine back to the county blacktop road, turned west a half mile, then north past picturesque, slightly rolling farmland and past two farms with large white barns and silver roofs. Another mile beyond he stopped partially sideways in the road and in front of a wide, wire gate spanning the entrance to a pasture. As he climbed down from the combine cab to open the gate, he heard a thump against the side of the grain tank. Looking back toward the combine, he saw a red-headed woodpecker fall to the ground.

"Geez, look at that," he exclaimed aloud to nobody in particular. He reached down beside the huge combine to pick up the small dying bird lying in the gravel. A drop of blood oozed from the base of its bill. For a moment he studied the bird in his hand, looked at its glistening red head and stark black and white body. Still holding the bird, he walked around to the opposite side of the combine and looked up. There it was. Some fifteen feet up in an electric pole he saw a neatly chiseled hole. He guessed he must have stopped right in the bird's favorite flight line to its nest. How on earth could the bird not have seen the combine? He looked again at the bird in his hand. The blood had spread slightly, and a damp stain appeared on its glistening red feathers. Zach felt its head go limp in his hands, the spark of life ebbing from its eyes. He laid it beneath the electric pole at the side of the road and walked back to the combine.

Del arrived minutes later with a truck and an empty gravity flow trailer. He unhooked the trailer at the edge of the first pasture, and left it for Zach to fill. Back at the homeplace, he grabbed a sandwich from the kitchen and climbed into the high cab of the semi he'd filled earlier in the afternoon. The big diesel rattled to life. There was enough time to take this load of fescue seed to Grand Forks and get the hauling fee too. Evenings, or when they had time, they hauled the seed directly to the buyer. The co-op in Prairie Point served as a local receiver for the seed company, but by taking it to Grand Forks the buyer would pay three-quarters of a cent a pound trucking fee—an extra two to three hundred dollars on each load. They'd spend half on fuel.

If there had been time, Del would have had the seed tested for moisture and purity at the local co-op, then have it rechecked in Grand Forks. If the machines weren't calibrated exactly alike, the values would be different, and he'd ask the buyer for the most favorable test. Without a competing moisture test, he knew the buyer in Grand Forks was unlikely to give him the benefit of the doubt, and it could make a difference of a hundred dollars or more on a load.

At the seed company in Grand Forks, a slender fellow in a cowboy hat and tight jeans took several samples of seed for testing. Moisture testing was automated, but testing for seed purity was surprisingly low-tech. The sample was visually inspected, weighed, and then shook through three separate small metal sieves, the impurities discarded each time. Finally, the remaining sample was placed in an air separator to blow off chaff and lighter weight seeds before being reweighed to arrive at pure seed weight. Impurities were running high this year, especially from Orchard grass seed, which couldn't be mechanically separated from fescue seed even by the seed company. Orchard grass was an alien from Europe. Like fescue, it was sometimes used in lawns and landscaping and also as a pasture grass, but it was aggressive and generally regarded as undesirable by fescue growers.

For the past six days, Del and Zach had harvested all day and trucked their fescue to the Grand Forks seed company at night, often not returning until nearly midnight. Tonight, he took Zach's partial load to nearby Prairie Point and was home by eight-thirty. It was only a few hundred pounds of seed, barely worth the time, but he'd be home early, and tomorrow fescue harvest would be over.

23

July 4
County Fair

July. Families enjoy Independence Day celebrations and parades in local towns; time-honored tradition of putting up hay begins in July with mowers, rakes, and balers humming in fields across the state; cattle are sprayed if summer flies become bothersome; blackberries and wild black cherries ripen; summer farm shows attract crowds; black-eyed susans bloom in roadsides; sweet corn ripens in gardens; birdsong subsides by middle of month; gayfeathers bloom on tallgrass prairies.

A woven wire fence enclosed three sides of a rectangular-shaped block of land where the Montgomerys' homeplace was located. Beyond that island of buildings, in the pasture to the east, Del kept a small herd of Brangus and Black Angus heifers. They were sleek and sturdy and would birth their first calves next winter. At night, a tall pole light held sway over a small sphere of the night around the house and buildings, but beyond the reach of the light in the dark, grassy pasture, where the air was cool and still and a hint of dampness crept in, you could hear the tugging sounds of the heifers as they pulled up mouthfuls of grass, and you could hear their feet rustling in the grass as they walked. Sometimes their hooves stamped heavily against the ground. As they grazed in the darkness, it was their rustling and chewing sounds, and occasional snorting breaths, more than their dark silhouettes that revealed their presence. In that same pasture crickets revealed their presence with "chirrup" sounds, sometimes in unison, sometimes falling silent, as did a sprinkling of fireflies who winked slowly, their low-wattage blinking seemingly random but never rising much above the height of the grass. High overhead, the stars of Hercules, the son of Zeus and the mortal woman Alcmene, seemed to blink too as he knelt, bow ready, as if perpetually completing his famous twelve labors. The stars that formed the figure of Hercules were modest as viewed from Earth, none bright enough to attract much attention. But they were, like the cattle and the crickets and fireflies and the sweet breath of evening that lay on the land, the essence of a midsummer's evening.

Summer was everywhere now, scattered along roadsides in the form of Amish selling vegetables and fruit, farmers trucking grain to elevators, machines in fields, and black-eyed susans and Queen Ann's lace filling roadbanks with color. The beginning of summer was marked by the solstice, but it was Independence Day, more than the solstice, that marked the moment when everyone was comfortably settled into the routine of summer. Across rural America's heartland, Independence weekend was a time of county fairs and celebrations, beauty queens, picnics, hot dogs, and children and parades.

The Montgomerys always made time for local fairs. Sometimes they attended two, even three celebrations in neighboring towns if their work was caught up. Zach would compete in a tractor pull, and there would be fireworks displays. Today they followed the county blacktop road west a dozen miles, past rectangular fields of knee-high soybeans, and fields of shoulder-high corn, and past green pastures with black cows and calves, and past roadside ditches with wildflowers. The road eventually turned north, and then west again into Ashton. Around Ashton, the farms showed an orderliness that was perhaps a reflection of the land, which was flatter and with larger fields than those near the Montgomerys' homeplace. It also may have reflected a desire of the mostly Germanic immigrants who had settled around Ashton to instill discipline and conformity on the land. Their white farmhouses were neatly arranged on little squares of land snipped from corners of forty- and eighty-acre fields of corn and soybeans. Houses, in predictable order, bordered roadsides, and around each house there was a cluster of mostly red buildings and idled farm equipment.

A year ago Amber entered Nathaniel in the Independence Day baby contest at the Ashton city fair. This year, at the age of two, he was riding with his father on a tractor in the parade. It was the first time Del and Zach had driven tractors in the Ashton parade. They took a pre-World War II vintage International, and another with a factor-equipped V-8 engine, unusual among farm tractors. In the parade, there would be other tractors, old cars, horses, wagons, motorcycles, bicycles, go-karts, people walking, some with their pets, others carrying banners, and all of them moving slowly along Main Street. A good crowd of people would be watching from folding chairs set up curbside.

The celebration provided a break from summer work. There was something comforting in the predictability of a local fair. There would be talk about contests and parades of the past, and there would be anticipation of what was to come, and then the fairs would be mostly the same each year anyway. But there was satisfaction in that sameness, in the anticipation that could be seen in the beaming faces of children as fair time grew near. And nothing, it seemed, so embodied the spirit of small town fairs and Independence Day celebrations as a parade.

Independence Day marked the onset of summer in another way. With the weather turning dry, as it often did at this time of year, farmers began making preparations for haying. For more than a week following the holiday, fleets of puffy white clouds floated across blue skies, temperatures soared, and farmers prepared for a prairie tradition that had been handed down to them from their fathers and grandfathers before them. They may have thought of it as just a normal part of summer work, rather than a tradition, but putting up hay for livestock for the coming winter had a history that stretched back centuries. Once it meant filling your hayloft with loose hay for the winter, and helping your neighbors fill theirs. Now it mostly meant fields populated with ubiquitous, oversized round bales that farmers lined up in rows at the edges of fields, or left cluttering fields until the dead of winter.

Harlan's job was to mow the hay fields and stay a day ahead of the hay rake and baler. What Harlan mowed today would be dry enough to rake and bale tomorrow. When farmers put up hay, they rarely mowed more than what they could bale the following day because unexpected rain would ruin cut hay. During haying season, a cloud seldom passed without scrutiny. Nobody wanted to get caught with hay down and a rainstorm brewing. Aside from losing the crop, there was pride in the act of putting up hay. Farmers didn't want their neighbors to see them screwing up a hay crop because of poor judgment about the weather. Once, after one of Del's neighbors mowed forty acres of hay, it rained before the hay dried. Then it rained again. After two rains and a week lying in the field, his ruined hay crop looked like decaying yard waste. Perhaps worse than the loss was the knowledge that all of his neighbors at the co-op and the local cafe would be talking about him. They would retell the story, recalling it months later with embellishments, perhaps, as they huddled over second and third cups of coffee and gossip on brittle winter days when they tried to remember the warmth of summer and the aroma of new-mowed hay on a midsummer evening.

The idea behind hay conditioning was getting it dried to a slightly crunchy feel, but not stiff and desiccated. There was a narrow window of time when hay conditions were perfect. Wait too long and nutrition would be lost. Bale too soon and it would mold and generate heat, sometimes starting spontaneous fires in barns. When that happened, it was the stuff of local legends.

For dinner Caitlin had fresh tomatoes and zucchini. "The garden looked dry this morning when I picked these tomatoes," she said to Del. "I'm going to have to start watering or we'll lose things. Pole beans are already dry."

"Well, your flowers look good," Del said with a grin. "That patch of zinnias is taking the heat. All the things you can't eat are doing fine, but all you can do is look at them."

"Won't hurt you to look at something pretty," Caitlin chided, "even if it doesn't make you money or you can't eat it."

Del glanced at Zach, then figured he ought to change the subject, so he asked Zach if he'd seen a highway patrol when he hauled a load of hay to his father-in-law. "That tractor-trailer's oversize load banner's a cop magnet. Especially on a hay truck."

"Yeah, they always want to see your permit," Zach agreed, "but I didn't see anybody."

"You know, I don't mind doing a favor for your father-in-law, but I sure ain't haulin' hay for a living. Those guys only get about two fifty to three dollars a mile to deliver hay. Can't make any money that way."

"Well, not if they don't load their trucks any better than we did," Zach said, laughing and remembering how he pulled onto the highway with the loaded truck the previous evening, and six big bales slid off the trailer and into the ditch.

"We didn't take enough time strapping the load on right," Del added defensively. "At least not all of it fell off."

"We might've been okay if it was old hay," Zach added. "Fresh bales are slippery."

Del had raked hay into windrows the day before, and Zach had followed with the baler. Harlan had continued mowing, but later came around to show Del where he'd mowed over a bumblebee nest.

"You don't have no cab on that little Farmall," he'd advised, "so you need to watch for them bees when you rake tomorrow. Just keep going. Give 'er hell when you go over the top of that nest 'cause they'll be coming out mad."

Del said he'd watch. Harlan didn't like bees, and the bees seemed to know it. Del thought maybe the bees could tell that Harlan was afraid of them. Amber complained that the only time she'd been stung was when Harlan stirred up a nest of honeybees in an old tree behind their house. He'd parked a noisy tractor right by the nest. When they got after him, he started running toward the house, swatting bees off his back with his cap. Amber was outside with Nathaniel, and one of them stung her. Del laughed when he heard what happened, but Harlan didn't see the humor in it.

Yesterday a roll pin in the 986's transmission broke. They towed the tractor to the shop. Del shrugged. Wouldn't cost much to fix, but it took time he didn't have now. He told Zach to use the newer Case-International Magnum for baling. Zach had wanted to use it on the baler, anyway, but in the past Del had resisted. He said the more working hours a tractor accumulated, the more chances there was that something would break. They could work on the magnum's engine, but it had a complex variable-speed transmission—essentially an automatic transmission with eighteen gears. It was maybe the only thing in Del's machinery inventory, trucks included, that he didn't feel competent to work on. If something went wrong with that transmission, the cost could be astronomical. He knew of twenty-five-thousand-dollar repair bills on tractor transmissions. Made him a little sick to even think about it because it could be more than the tractor was worth. He'd seen newer tractors than his, big ones, on sale lots being sold for parts—transmissions were too expensive to fix, but today he had to make an exception.

"Anyway," he rationalized as he talked over the decision with Zach, "that Magnum has a reliable history. I never heard of a single one that had to have major transmission work. I just don't want to be the first."

24

July 24
Prairie Tradition

America's midwestern farm belt, with its crisp, angular fields and endless rows of soybeans and corn, owes its existence to sod-busting pioneers that ploughed out almost all of the vast tallgrass prairie. Today it's the nation's breadbasket. A century and a half ago, it was a two-dimensional landscape of grass and sun stretching from Manitoba to Texas and east to Indiana—more than 170 million acres. It wasn't all grass. A few trees speckled the region even then, mostly in lines along streams, or in precarious little tree islands that somehow survived prairie fires. Eastward there was an ever-shifting boundary of pioneering trees where prairies ceded ground to eastern hardwood forest. Today the tallgrass prairie is mostly gone; less than four percent of it remains, mostly in northern Oklahoma and the flint hills of central Kansas.

Efforts to break the tough tallgrass prairie began prior to the 1830s, with farmers using horses to pull walking ploughs, but their success was limited until the introduction of the self-scouring steel plow and later, after the Civil War, steam-powered tractors. Unlike horse-drawn implements, these slow-moving, hissing, steam-powered behemoths were able to pull large gangs of steel ploughs and break even the toughest prairie sod. Tallgrass prairie root systems that extended twelve to fifteen feet below the soil surface and ensured the survival of grasses against the hottest and driest summers and coldest winters were no match for steel ploughs. In a few decades, a biome that once seemed endless was gone forever. Unlike most forests, which will regenerate over time if remnant woodlots remain, prairies cannot be restored to anything resembling their original form in a timeframe meaningful to humans.

By the early 1900s, most prairie land suitable for agriculture had already succumbed to the plough. Missouri once had an estimated twelve million acres of tallgrass prairie, almost twenty-eight percent of the state. Today less than one half of one percent remains, most of it in the southwestern part of the state. North of the Missouri River, where rich glaciated soils predominate, there is almost none. A few tracts of native prairie have been acquired by the state or are owned by private conservation organizations. The rest is in private hands—mostly little twenty-, forty-, and eighty-acre prairies, rarely much larger, tucked away on farms and surviving in places where the soil is too rocky or sandy to farm. Prairies mowed for hay each summer are usually free of woody growth. Other prairies languish, damaged by overgrazing, or are overseeded or invaded by non-native grasses, which tend to crowd out native grasses. In the spring and early summer, these little prairie fragments form checkerboards of green rectangles pixelated with reds, yellows, and blues of paintbrushes, coreopsis, wood betonies, ragworts, phlox, and violets, and are scattered across southwestern Missouri as colorful reminders of this once-magnificent grassland. Soon Del and Zach would be putting up prairie hay on some of these remnants.

Zach, in old jeans and greasy red T-shirt, sat on the ground in front of a rotary drum mower. It was an improbable-looking machine, brimming with gears and belts and spinning disks. At its most basic, it was not unlike several bulked-up weed-eaters linked together and cloaked in canvas. A power take-off shaft at the rear of a tractor provided power. These mowers were the crowning achievement of a century and a half of agricultural engineering's obsession with finding better ways to cut grass, and these were light years from a scythe, or even the sickle mower Zach's grandfather used. Suspended beneath the heavy ten-and-a-half-foot steel supporting beam were a half-dozen saucer-shaped spinning disks, each a little bigger than an extra-large pizza, and each with three free-swinging knives. A durable tent-like hood draped over the superstructure shielded operators and naive bystanders from debris that would be mere batting practice for the machine's blades. Zach raised the mower's canvas hood and began sharpening the stubby knives with a hand-held electric grinder.

"How's it look?" Del inquired, walking over to examine the blades.

"Not bad. I need to change one or two knives," Zach observed, running a finger over a worn edge. "Soon as I finish, I'll grease it and we'll be ready."

"I just greased and serviced the baler," Del added. "I'm going to the co-op to get calf supplement and cattle cubes. We need more baler twine too.

What do you think? Twelve balls of sisal and four of that plastic polypropylene stuff? We'll sell the hay on the seventy acres we're baling now. Four balls of poly twine should finish that. We'll wrap the rest, what we use ourselves, with sisal."

"Sounds right," Zach said as a shower of fiery orange sparks erupted in spurts from his portable grinder.

Pausing for a moment, he said, "Could you bring me a set of metric sockets? I gotta have metric to change the knives."

Del turned toward the shop, almost tripping over Boy the shop cat, who had padded up silently behind him. Some farm machinery was metric now, especially that out of Europe. It was a nuisance having to use two sets of tools and remembering to keep both in their truck for field use. "The hardest part," Del groused, "is not mixing them up."

Before Del left, a neighbor, Bobbie Ray, drove his new four-wheel drive Case-International tractor into the driveway and stopped beside Del's shop. His wife Joanne followed in a pickup. Bobbie Ray's knees were bad, and Del could see him grimace as he eased himself down the narrow steps from the tractor's high, glass-enclosed cab. He left the rattling diesel engine running. With a wave of his hand, he motioned for Del as he moved stiffly around to the front loader on the tractor. Del had known Bobbie Ray almost all his life. He was a good neighbor. A lifetime of heavy work had worn down his once-strong body. Still, he refused to quit farming, and, in this, he was like most farmers. It was a life that they could not imagine leaving. Bobbie Ray wore a red seed company cap and bib overalls partly unbuttoned at the sides. His face was deeply tanned. Removing his cap, he wiped his forehead with a sleeve and began talking in a loud voice in order to be heard above the batter of the idling diesel. He didn't like the automatic self-leveling devise on his tractor's front bucket loader. Could Del or Zach disable it? Over the noise, both men pointed to make themselves heard as they leaned in close to inspect the loader's hydraulics, welds, and mechanics. Bobbie Ray walked around to the tractor cab and, without climbing up into it, reached up and pushed a lever that raised and lowered the bucket to demonstrate what he considered the problem.

Del, struggling to hear, said he thought he could fix it later. "I got hay down today. Can you bring it back in a few days?"

Bobbie Ray nodded, then looked up and saw his wife Joanne inside the tractor cab. He seemed grateful that he didn't have to get back up in the tractor seat and walked back to the truck. Even before he reached the truck, she had whirled the tractor around in the driveway and was on her way home.

"Wonder why he wants to disconnect the self-leveler?" Del said, turning to Zach, who was still sharpening rotary mower blades. "I always thought that would be a nice feature to have."

"Don't know," Zach replied. "It seems a shame to cut and reweld on that new loader."

Later in the morning, in Prairie Point, Del backed his black flatbed truck up to the co-op's loading dock. He climbed the concrete steps to dock level and walked inside, winding his way between mountainous stacks of sacked Cattle Charge, Beef Feeder, Stock Feeder, Vit-A-Zine, Cattle Cubes, Easykeeper Horse Feed, Muscle Pig Corn Chops, Soy Hull Pellets, Calf Finisher, block salt, pelletized salt, poultry grower, scratch grain, soybean meal, Hunter's Special dog chow, Black Gold cat food, sunflower seeds for birds, and huge stacks of baler twine of various sizes. The warehouse fairly tingled with pungent odors of dust and feed and chemicals. Moving past the piles of feed sacks, some rising well above head height, Del suddenly felt small, like he was moving through a landscape hewed by giants. He stopped briefly to tell Riley, the warehouse man, he'd need a dozen bags each of calf supplement and range cubes, and then headed through a two-way swinging door to the front office to get his order ticket.

Del needed sisal baler twine, too, but the manufacturer was now offering it in two spool sizes, and he wasn't sure if his baler could handle the larger spools. The girl at the front desk, a pert high school student in brown curls and tight jeans, already knew which spool size would fit when Del told her the model number of his baler. He looked surprised, then even more so when he discovered she knew the make and model of every baler of nearly every farmer in the area and exactly what size twine spool fit each one. He questioned her about the longer-length spools, and she pointed out to him that Jack Larousse had tried some of the longer spools and they wouldn't fit in his baler. "They probably won't work in yours, either," she said, beaming.

"Thanks," Del said with a bemused grin that revealed a newfound respect for the co-op's newest employee, then turned and walked back to the antiquated soft drink dispenser in the corner of the office and fed two quarters into it while glancing at a grease board listing current prices of a dozen kinds of cat and dog food. A Diet Coke rattled noisily down the slot and into the dispenser below. By the time he got back to the warehouse, Riley had the supplement and range cubes and twine loaded.

It was ten o'clock. A light southwest breeze had burned off the morning dew from the previous day's new-mowed hay. The sky was hazy, a frosted bulb with bluish tint, already hot, a good day for putting up hay. At the house, Del and Zach filled water jugs, adding several trays of ice, and left for the hay field three miles to the west. On the way they passed Harlan, who was clearing recently baled hay from a field. With each trip his tractor ferried two giant bales, one impaled on the loader up front, another on a spike attached at the rear. He'd already cleared a third of the field, lining up the bales at the side of the field. The newly cleared part of the field looked stark, shorn like a barbered crewcut.

"As usual, he was thirty minutes late getting here this morning," Del said, "but it looks like we're getting our money's worth out of him now."

Most of the hay they were baling was fescue. Missouri farmers had been so eager to plant fescue for its seed and winter forage that from the seed in a single field in Kentucky in 1931, there was now an estimated seventeen million acres growing in Missouri. Fescue was also used for a hay crop but had a reputation as poor quality hay, although that had more to do with the haymaker than the hay. Like most cool-season grasses, it was nutritious if cut early, at the onset of seeding, but less so in mid-summer when temperatures were hot and growth slowed. Some farmers scoffed at late-baled fescue hay, calling it a "gut stuffer." Del usually baled it in early summer when nutrition was declining, but he offset the reduced nutrition with cattle supplements.

Early settlers and farmers relied on native prairie grasses for haymaking. Native grasses grew best in summer heat and were known as warm season grasses by agronomists. Superbly adapted to the stresses of drought and fire and insects, they required no maintenance, didn't ask for fertilizers or pesticides or herbicides, nor did they require irrigation or reseeding. However, unless prairies were mowed and baled each year, they needed periodic fires to prevent invading wood growth, and burning was increasingly regulated and difficult to undertake. Over time many native prairies were simply replaced by monocultures of higher-producing but higher-maintenance grasses. More than any other grass, it was fescue that replaced these remaining prairies in Missouri.

It was a reshaping that cattlemen approved of, and one that biologists didn't necessarily like. Cattlemen like Del argued that fescue controlled erosion, and provided seed income and winter grazing. Biologists argued that fescue crowded out native grasses and was too dense for wildlife. Fescue also requires annual doses of nitrogen, has to be replanted after a few years, and most of it in Missouri was infected with a fungal endophyte that could affect cattle health. Both parties were right. Native grasses went dormant during the long winter. Fescue grass didn't, which was an advantage to cattlemen. Fescue developed a sod so dense it was effective in controlling erosion, but the dense sod limited plant diversity and the variety of insect foods that wildlife needed. It also was poor wildlife cover because the sod didn't have the spaces that birds and animals needed when seeking cover. It came down to a question of managing for quails or cows. Short term, fescue was the economic winner. Longer term, fescue might illustrate the risk of dependence on a monoculture. Its final economic and environmental chapters have yet to be written.

The hayfield was dry. In places the short fescue stubble was yellowish and spread in irregular straw-colored patches across the field where it mixed with greener areas that held moisture. Del loaded the baler with twine, tying the four spools together like an electrician wiring a circuit in series so when the first spool was used up the next one below would be used and so on. Zach had raked two long windrows by the time Del entered the field, engaged the power-take-off to apply power to the baler, and revved the tractor engine to set a ground speed of a little over five mph. He might have driven faster, but the field was rough, a hail of hoof prints where they'd fed cattle in last winter's mud.

In those few moments, Del put into motion an institution in American farming, a prairie tradition of making hay that was a deeply ingrained image in American agriculture, but one dramatically changed in recent decades. Images of hay crews, hay wagons trailing behind balers, men and boys stacking hay in lofts in hot, dusty barns on summer afternoons, and clattering balers endlessly churning out little square bales was standard post-World War II procedure. That image has largely disappeared, although small square bales, as well as large square ones, still predominate for high-quality hay such as alfalfa, partly because it can be stacked and shipped more efficiently to dairy producers, horse owners, and to overseas Asian markets.

For millennia, haying was as simple as a load of grass on someone's back. Mechanical balers, introduced in the late 1800s, set in motion a revolution in an agriculture tradition that had remained unchanged almost since humans first domesticated livestock. For the first time in history, humans had the ability to transport large quantities of forage crop and to better preserve the nutrition in it. Early balers were stationary presses anchored to the ground. Hay or straw was forked into the top of the press and a plunger, powered by horses walking in a circle around the press, provided the compression to form small rectangular bales. Stationary hay presses were slow, inefficient, and required large crews—up to a dozen farmhands—half to feed hay into the press, tie bales, remove bales, and tend horses, the other half to mow, rake, and transport loose hay to the baler, all of which required almost as many horses or mules as men. Even with the use of gasoline-powered engines, stationary baling remained extraordinarily labor-intensive. It was a quantum leap over loose hay, but the reign of stationary balers was mercifully short-lived.

By the end of World War II, tractor-pulled balers capable of picking up hay from a windrow and automatically tying the bales gained rapid acceptance despite a cussed reputation for malfunctions. And these new balers still didn't eliminate the work of storage. Square bales, which were really rectangular in length, didn't shed water and had to be moved to storage quickly before they were ruined by rain. To counter this storage problem, roto-balers, which produced small, cylindrical bales, were introduced and became popular in the 1950s and 1960s. These round bales shed water and fared much better than square bales when left outside, but the balers were prone to mechanical problems and dangerous to careless operators. Through it all, haying remained hot, dirty, backbreaking work. The amount of human labor involved in it didn't change significantly until the 1970s, when giant bales, tractors with air-conditioned cabs, and hydraulic loaders spread across America's heartland. After that, an entire hay crop could be put up—mowed, raked, baled, hauled, and fed—by just one or two people, and the hay never needed to be touched by human hands. It was indicative of how quickly American agriculture was changing, and of how quickly backbreaking manual labor was being replaced by technology. As horsepower, hydraulics, and technology ramped up on farms, so did the exodus of humans—a once enormous pool of manual labor—from farms across the country.

The baler Del used was ten years old, but new enough to employ several state-of-the-art features. An electronic monitor inside the tractor's air-conditioned cab displayed nearly every imaginable baler function but still provided for manual override. Lighted arrows on a monitor told an operator where to drive so the bale wouldn't be lopsided, and malfunctions were signaled with warning sounds and lighted messages. Del only infrequently looked back at the baler. He needed only to watch the monitor and drive so the tractor perfectly straddled the windrow.

A modern baler looks like nothing so much as a huge mechanical insect pausing to oviposit immense cylindrical eggs as it moves across a hay field. The inner workings of a hay baler like this, one that spins windrows of loose hay into almost perfectly formed half-ton to one-ton cylinders of tightly wrapped hay, are elegantly simple if not initially intuitive. The mystery in them is, perhaps, not that a bale can be formed, but that a bale is ever initiated inside of the baler in the first place. Beneath the baler, a rotating metal drum with springy steel tines deftly picks up the windrow of hay much like a pick-up attachment on a combine picks up windrowed stems of fescue. The baler feeds the loose hay almost directly upward between two sets of flat rubberized belts that are so close together they almost touch, one set moving upward in a clockwise direction, the other set moving downward counter to the first set. Hay rapidly forced up between the belts immediately begins to rotate between them, being tugged in two directions simultaneously. As more hay feeds into the machine, the incipient mass of hay quickly grows into a tight-spinning cylinder of hay that forces the two sets of belts, which are under high tension, to separate. The gestation period of a bale inside a baler is only a minute or two—dependent mainly on the quantity of hay in the windrow and how fast the operator drives. Once optimal bale size is reached, the baler begins making decisions on its own. The driver is signaled to stop, which halts the inflow of hay. A twine arm descends and wraps twine around the spinning bale. Once wrapped, a blade severs the twine, and the entire rear half of the baler unhinges at the bottom and opens wide like the jaws of an enormous beast, allowing the bale to be pushed rearward beyond the baler.

A hundred yards down the windrow, the monitor beeped, signaling "Full Bale." Del hit the tractor's clutch and brake simultaneously, bringing tractor and baler to a full stop while the baler wrapped the bale with twine. It took about twenty-five seconds. Del's input consisted of touching

a pressure-sensitive button in the upper-right corner of the monitor to initiate the twine-wrapping process. The rest was automatic. Brief monitor messages indicated "Twine Arm Down," and "Gate Up," and "Ejector Out" and so on. Once the finished bale was birthed, the monitor flashed "Drive" and a new cycle started. Some farmers wrapped bales with a cellophane-like plastic wrap rather than twine. The wrap reduced spoilage and made long-distance transport easier. It also was much quicker than wrapping with twine, requiring a bale to rotate only a revolution and a half. Initially, plastic-wrap was expensive and used mainly on hay destined for long-distance transport, but the price is now less than that of sisal and is widely used.

In an hour and a half, Del baled a third of the twenty-acre field. In that same time, Zach windrowed more than half the field, pulling a star-wheel rake behind a small tractor. The rake was an odd-looking, V-shaped implement that employed five giant wheels on each side. Each wheel looked like a five-foot-high incarnation of a child's drawing of the sun. Powered only by the drag of friction against the ground, the big wheels deftly swished mowed hay into an endless breaking wave that moved toward the center of the V. Where the waves met, they left the hay stems essentially standing on their ends in a perfectly formed double windrow.

Shortly before noon, Del phoned Zach. "We better knock off for our dinner. Caitlin will be home from the post office soon. She's got to be back at work by 12:30. I'll get the truck. I'm closer to it than you are."

On their way home, they picked up Harlan. He was still hauling bales. He didn't stay for his midday meal, instead he headed for his truck and boat parked by the grain bins. He'd already told Del he needed the afternoon off.

Del briefly talked with him. He knew Harlan was upset about his marital problems. For several nights he'd slept in his boat or with a friend. His wife liked to drink; he liked to fish. The combination didn't seem to mix. Del figured this time it might be different. Harlan had finally moved out.

"A few weeks ago I had to spend three days in the same clothes 'cause I couldn't get in my own damn house," Harlan railed. "Finally a sheriff's deputy gave me ten minutes to get my clothes. Ten whole goddamn minutes! Can you believe it? Just to get in my own damn house. I can't take it no more."

Harlan gunned his pickup out of the driveway, boat and trailer bouncing behind. The boat was nearly new, beautifully contoured and with a glittering blue and silver coating, one of the few things of value that Harlan owned. It may have been the only thing that was a source of pure pleasure to him.

Del and Zach, as usual, talked corn and hay crops and weather over their midday meal. Caitlin put out cold cuts, meat, potato salad, and tomatoes, and left for work before they finished.

"Corn's looking good," Zach said. "Ears so long and heavy they're bending over. Last year was so dry the ears never filled out. Not heavy enough to lay over."

"We're pretty much out of the woods with the corn crop. It's almost made now, even if it does turn dry. A few years ago about his time I was at the FSA office signing up for disaster relief payments from the government," Del added.

"So was everybody else. Corn prices were still low then. By the way, you planning on mowing on Sunday?"

"Yeah, we'll bale on Monday. Hay's curing fast. Hardly need a full day in this sun."

"If Harlan comes back to mow on Monday and Tuesday, then we ought to finish by Wednesday."

"But then we got about two weeks' worth of hay hauling to do. We got over twelve hundred big bales already. That's two hundred more than usual. Harlan's barely started moving them out of the fields."

They figured the weather would stay good now. It was almost August, typically a dry month. Even if they lost a little mowed hay to rain, it wasn't as bad as dry weather burning up summer crops. They had plenty of hay, but a lost crop couldn't be replaced.

Del and Zach finished baling by five in the afternoon, early enough to work in the shop for an hour or two. Zach had stripped some gears in his pulling tractor over the weekend. If he missed another pull, he would lose his hold on first place in point standings. He also wanted to work on his new "super tractor." He was still hoping to get it ready for a trial run before the end of the pulling season.

Late in the evening Harlan phoned. Del and Zach were still in the shop with Zach's 460 tractor transmission spread out across the shop floor.

"How you doing?" Del asked.

"Not worth a damn!"

"How's that?"

"Judge said we need a cooling-off period. Me and Judd is fishing. Best thing to do is get away from it all."

"Well, try to enjoy your weekend. We'll see you Monday."

"That Harlan?" Zach asked, knowing it was.

"Yep. Judge's decision don't seem right to me, but I only heard Harlan's side of the story."

There had been more damage to the transmission than Zach suspected. He thought he'd chipped some teeth off fourth gear. Once he got inside the transmission, he saw broken teeth on two other gears. He also saw that the teeth on the bull gear wheels were heavily worn on the inside. He'd swap sides with them so they'd wear on the opposite sides. That would likely be the most time-consuming part of the repair, and he had to be careful to avoid disturbing the axle bearings during the move or a leak could develop. He'd have to remove both rear tires to finish.

Zach had all the replacement parts for the 460's transmission and differential, and the reassembly went quickly. He finished most of the work Saturday evening and spent part of Sunday morning mowing. It was one of those rare times when they were almost caught up with farm work—enough that everyone decided to go to an antique and steam-powered tractor show in the afternoon—a diversion after a long week.

It was hot and sunny, and they arrived too late for the threshing demonstration.

"Just as well," Del said. "I've seen those things enough times anyway. Glad I never had to harvest that way. Awful hot working around all that steam and heat all day. Took an army of men to get anything done and nearly as many women to keep 'em fed. It's a wonder anybody got any farming done in those times."

Zach was less interested in the old steam-powered equipment than Del, a generational thing, he figured, and headed for a display of older gasoline-powered tractors. Caitlin and Amber and Nathaniel looked for a shade tree.

Steam-power shows were held every summer across the Midwest, but attendance was low, mainly older farmers, and only a few of them could even remember when American agriculture ran on steam power. They were children then. Now they were old men trying to keep memories alive. Some remembered it as the "good old days." Others said it was just as well those days were gone. They remembered sweaty horses and barns full of fleas and binding shocks of grain by hand and threshing crews sweating under a hot summer sun, and grain scooped by hand, and itchy chaff from threshing machines. In that dawning of American agriculture, the work was back-breaking: planting crops, cultivating with primitive horse-powered implements, cleaning manure from barns by hand, slow and inefficient hay presses, bull rakes, and hot, dusty haylofts. On many hot summer afternoons, it had taken more than a little homemade beer and wine to get hay crops in the barn loft.

Later, as they were leaving, Del said to Zach, "You know, the biggest loss in American agriculture isn't these old machines. They were doomed anyway. What's been lost is the sense of community that went with those early days."

"Hard for me to imagine what it was like," Zach said.

"It was different then. Neighbors helped each other. Harvest and haying crews worked and ate together. They shared something that we don't have now. Same with the women. The wives cooked together and took care of their men. People had a sense of place in those communities. That's slipping away. Already gone, really."

Zach had heard his dad talk about this before. He'd grown up farming but had never experienced that kind of sharing of farm work within a community. Everybody worked alone now. Machinery was bigger, more efficient. He and his dad had never needed the kind of farm help that was required in the past.

When Del was a kid, a few farmers still thrashed grain using the old methods, but even then they used gasoline tractors, not steam tractors. He'd experienced just a little of that earlier era when farming was mostly about trying to feed your family and about growing your kids and sharing with friends and community. Today it was about bottom lines, bioengineered seeds, herbicides, complex farm equipment costing hundreds of thousands of dollars, computers, government regulations, and support payments. The work wasn't as physically hard, but it was a lot more complex and there was more to worry about.

Zach listened as Del continued. "I think a lot of these old guys come to these shows to remember when farming was about something else, but when they look at these old machines they're not really thinking about them. They're nothing to get nostalgic about, just hot and dangerous, and these guys have forgot all the hard work. What they're really remembering is the way it was with neighbors and communities."

25

August 15 & 18
State Fair

August. Farmers move hay bales out of fields and to storage sites; winter manure in corrals and barns is hauled to fields; pasture mowing continues throughout month; family vacations planned; farm families attend state fairs, show cattle, livestock, food, and domestic products and compete in state-wide tractor pulls; early-planted corn harvested by end of month; purple-martins gather for migration; wild grapes and elderberries ripen; dry-weather cicadas call incessantly; long, hot days often spell the onset of a summer drought.

"Did he see me pull?" Amber asked, breathless with excitement as she rushed back to the pavilion bleachers where Caitlin waited with Nathaniel.

Caitlin shook her head. "No, he fell asleep before they announced your name."

Amber's face fell, showing her disappointment as she reached out to take Nathaniel from Caitlin's arms. "Sleepy boy," she said, rocking him from side to side.

"He's been here since ten this morning," Caitlin said. "Stayed awake over four hours. Looks like he's going to miss his daddy's pull too."

"Maybe he'll wake up. Zach doesn't pull for another two hours."

Amber was so excited she could hardly sit still. Driving her old Farmall H tractor, the one Harlan and Zach had switched motors in last winter, she'd just bested twenty-nine entries in the antique and classic tractor pulling contest at the state fair. She loved competing, especially when she could beat out some male drivers, but she never expected to take first place. She'd done it by pulling just a few inches farther than her nearest competitor. There wasn't any prize money, but she would get a monster trophy, a gleaming four footer with a little tractor emblem on the top.

"Too bad the tractor doesn't have a woman driver," she said. "I ought to complain to them about that."

Del and Caitlin had always gone to the state fair, even before they had children. When their daughter Wendy was in grade school, Caitlin helped her show cattle, so the fair became an annual family outing. Zach also had helped his older sister show her cattle for a few years but then got more interested in machinery.

Amber hadn't gone to fairs when she was young. "We moved quite a bit," she told Caitlin, "and with my parents not being in farming, nobody in the family had much interest in them."

"It was the opposite with us," Caitlin said. "We always went to fairs. When the kids were younger, we'd unload Wendy's cows and calves at the cattle pavilion. Then we'd get Zach entered in a tractor pull. The pulls used to be one-day affairs. But if you showed cattle you had to keep them at the fair for four days."

"You stayed at the fair four days?" Amber asked in disbelief.

"Back then it was a requirement that show cattle be on public display for at least four days. They don't require that now, and if people bring cattle it's usually just for a day, maybe two. You can tell. There aren't nearly as many cattle on display now. Hardly worth walking through the barns."

"But what did you do for all those days?"

"Somebody had to be there to feed and water them and keep them brushed and groomed. It was a lot of work."

"You mean you had to wash them and brush them?"

"Oh, yes. And clean up after them too. If you showed for judging, the grooming was a big job. There was a place to wash them, and we curried their coats and brushed their tails. Even washed and cleaned their hooves. We always showed Angus or Brangus, and they have short hair, so they were easier. Wendy would clip any long hair around their head. They looked nice that way. The long-haired breeds, like Herefords, were a lot more work. You had to spend hours brushing them. Some people would bring hair dryers to dry and curl the cattle's hair."

Amber's eyes widened. "I didn't know you had to do all that."

"Wendy loved to prep her cattle for judging. When she was in a high school agriculture class, Del helped her build a portable holding chute for her cattle where she could confine them for washing. It helped them get used to it so they would be calm. You know, even this year, with her new baby and working full time, she's planning to take off a day to go to the state fair. She's been away from the farm for years, but the farm's still in her blood."

Amber wrinkled her nose. "I always thought the cattle and hog barns didn't smell good."

Caitlin laughed. "Maybe to some people, but Wendy grew up with cattle. When she was in school, she'd feed her cows and calves every morning and put a halter on them and lead them around, even in the winter. That's how they got so gentle. She loved being around cattle. She was always so excited when she walked into the arena to show."

Amber looked surprised. She thought she knew Wendy pretty well, but all of this was news to her. "I was never around cattle until Zach and I got married. I still like it better when there's a fence between me and the cows."

This year the Montgomerys went to the fair twice. On Sunday Zach entered his Farmall 460 in a pulling contest, and on Wednesday Amber entered the classic and antique pull. A brief shower delayed the Sunday pull, and Zach drew the first pull after the rain. The dirt track was slippery, and the first few pullers didn't do well. After that conditions improved, and later entrants fared better. Zach felt lucky to hold on to third place, but with the rain delay the pull extended well into the night. Then Del couldn't get his truck and trailer out of the parking area because the driveways were blocked, and they waited over an hour before they could move. It was nearly two a.m. before they got home. Amber was exhausted. She had to be at work by eight the next morning and didn't want to go back to the fair on Wednesday. Then, at work on Tuesday, a high cabinet drawer in the doctor's office where she worked fell and hit her head. It knocked her to the floor, and the doctor said she might have a mild concussion. It was serious enough to warrant an X-ray, and she was sent home to rest.

Both families had planned to leave again early Wednesday for the state fair, but Amber and Zach overslept. Del called at 6:30 a.m. and woke them. Zach hurried over to the homeplace and left with his dad. They took Amber's tractor and registered for her pull, scheduled to start later in the day. Amber had a headache. She worried about the accident as she dressed Nathaniel and drove to the homeplace to eat breakfast with Caitlin. The three of them didn't reach the fairgrounds until late morning. That afternoon, after winning a trophy, her accident the day before seemed like ancient history.

The appeal of state fairs runs deep in rural America, but fairs have changed so much since their humble beginnings more than two hundred years ago that fairs today bear faint resemblance to those of early years. However, fairs have always been an important venue for rural families to display their products and creations, and to learn what their neighbors were doing. In early rural environments, fairs were a valuable outlet, especially for farmers' wives who were often isolated with few social outlets beyond local communities and church. Fairs, with their horses, bands, carnivals, music, food, crafts, and recipes, brought an infusion of worldliness to isolated rural families.

Today, in a world of rapid transportation and instant communication, modern fairs play a diminished role in the lives of farm families. To stay relevant, fairs, especially at the state level, have undergone dramatic changes. Early fairs featured livestock exhibits, threshing-contests, and skills essential in early America, such as blacksmithing, butter churning, quilt making, even tintype photography. These venues have almost all disappeared from modern fairs. Agricultural equipment and technology, once staples at state fairs, have mostly migrated to specialized trade shows. Gone also are stunt flying, airplane rides, motorcycle barrel riding, and other events banned because they were deemed dangerous. Auto racing now overshadows harness racing, and horse- and pony-pulling contests have been replaced by truck- and tractor-pulling events, such as those in which the Montgomerys were participating.

Still, some aspects of state fairs endure—cooking contests, floriculture and flower arranging, home economics, fine arts, foods, and dairy. Show animals including horses, beef cattle, swine, sheep, poultry, even goats, rabbits, and llamas are still proudly entered in some state fairs and compete for prizes in adult and youth categories although in smaller numbers now. Fair events today are apt to include many new entrants, such as backyard chefs, weightlifters, musicians, even chain-sawing, yet a few novelty throwbacks to earlier eras remain, such as horseshoe pitching, bale-throwing, hay moving, egg-gathering, and gate hanging. The array of categories in which contestants can show cattle now includes grand champions, reserve grand champions, junior champions, and so on, giving almost everyone a chance to compete.

No state fair would be complete without a carnival, nor such staples as corn dogs and cotton candy. Entertainment at state fair carnivals remains focused on food and fun, although events and rides today bear scant resemblance to fairs of even a couple of generations earlier. In a world of loud music, bright lights, stuffed animal prizes, carousels, and carnival rides, the fast-talking hucksters still make their pitch to crowds, but the innocence of past eras has been replaced by a worldliness and sophistication today that did not exist in earlier years. Fair goers today no longer live isolated rural lives and are likely to have less need for what fairs have traditionally offered.

"What's that I smell?" Del asked. "It's like something's burning."

"I smell it, too, but I don't see anything burning," Caitlin said, looking around at the fairground parking lot and the long rows of parked cars and trucks that surrounded them.

"Let me look at that propane grill of Zach's," Del said, climbing up onto the wooden bed of the trailer.

A thin line of smoke seeped from beneath the grill. He scooted the grill to one side with his boot. The legs of the grill were folded under on one side. He could see the hot grill touching the wooden floor of the trailer. A small cavity had burned into the wood of the trailer.

"Hey!" he exclaimed. "You guys are burning a hole in my trailer! We got to move this thing quick! Help me get the food off this grill."

Zach, resting in the shade of their screened tent, heard the commotion and jumped to his feet. Outside, he looked at the hole, then with studied casualness, said, "Oh, it's just a little hole. Not like the whole trailer's on fire."

Caitlin looked chagrined. She'd set up the stove with Zach's help. The trailer was on an incline, and to level the stove they'd left one set of legs unfolded. She guessed the wooden trailer bed got scorched when they were preparing dinner and then charred the wood while they cooked their supper.

It was the first time they'd brought food to prepare at the state fair, but the tractor pulls were longer now, so they had time to cook. Even in the past, when Caitlin and Wendy showed cattle, they hadn't cooked, just eaten sandwiches and cold cuts to keep things simple. They brought cots and sleeping bags and the two of them slept beside their cattle trailer, or near the cattle, usually for three or four nights. Caitlin missed that.

"Sure, the days were long," she told Amber, "but I enjoyed the sounds of the fair, especially at night. The air was cool, and you could hear people talking as they passed in the evening. There was music, and the sounds of the carnival were always in the distance, and you could hear the animals as they moved in their stalls during the night. It was all so exciting."

"Maybe it was," Del ventured, "but showing cattle sure was a heck of a lot of work." Yet he knew Caitlin and Wendy wouldn't have traded those days and nights for anything. Rural America ran in their veins.

26

August 25
Summer Haze

Del said farming was mostly just hard work, but there were times when there was a lot more work. Those times usually coincided with work most emblematic of farm life—spring planting, summer harvests, putting up hay, autumn harvests, and cattle sales. During these periods, farmers often had time for little else. Most farmers liked what they did, even liked the hard work, although few would admit it. There wasn't a lot of room in their lives for long weekends off or vacations, but there were slack periods when farmers and their families could relax. Farm wives, more than their husbands, might long for vacations or time away from the farm, and younger rural families were more likely to take time off than their parents and grandparents. Still, farmers like Del, who had cattle and did row crop and grain farming, found it difficult to get away for more than a few days. A vacation, if it was on the horizon, had to be squeezed in between farm work.

Del and Caitlin tried to set aside a week or ten days for a summer vacation. Over the years they'd visited many parts of the United States. A trip to New York City seemed daunting, but after spending several days there, Del said the hardest thing about it was finding a taxi driver that spoke English. He and Caitlin had talked about farm tours to Brazil or Argentina or Europe, but scheduling a trip in advance had been difficult because of the risk of weather interfering with harvests. During the last few years, Caitlin lobbied for shorter trips to sites where she had access to county court records to trace genealogies. She had followed her family's records back to France and England. Del's family had proved more difficult. After following it back a few generations to Kentucky, the trail had gone cold.

The middle of August was an in-between period when farm work slowed, a brief interlude after summer harvests and haying and before fall harvests began. August usually meant long, hot days and sleeping-on-the-porch nights. Cicadas droned through the heat of the day, cattle took to standing in creeks and ponds to cool themselves, and grasshoppers were thick in weedy places. The growing season was in decline. Gardens and fields slipped into quiet senescence in August, weary of summer-long battles with heat and insects and disease. It was a time when work schedules were sometimes derelict as rural families made time for state fairs and church ice-cream suppers and vacations. Del remembered his grandfather saying that during the summer, families would often spend a few days, even a week or more, camping and fishing along rivers that could be reached in a day or two with a team of horses and a box wagon. Trips of thirty or forty miles were grand adventures.

For most of the past two weeks, it had been hot and dry. Summer droughts were something midwestern farmers dreaded but had learned to expect. They grumbled if it happened, and if it didn't they grumbled that it might happen. Fields once vibrant and green were reduced to muted palettes of grays and browns. Pastures turned dull and dusty. Even the blues of chicory and yellows of waist-high sunflowers in roadsides seemed languid and diminished, their blossoms nodding in the oppressive heat. This morning a dull haze hung in the air. The sun was a white-hot sphere spreading a withering glare across parched fields. A malaise clung to the land. Not a bird called. Even the cicadas were still. Listless cattle moved slowly in fields, the heat sapping their strength. They moved, not purposefully but of necessity, between drying ponds and islands of shade beneath scattered trees. Del saw the transforming effect of a succession of hot, rainless days on his crops. He felt fortunate that the heat had come late.

"Won't hurt the corn much now," he told Zach. "It's already almost made. Early-planted beans are pretty far along, too, but late-planted beans will get hurt if this lasts."

The kernels of corn were hard and flinty, already dented and almost ready for harvest. Once crops got to that point, the drought would have relatively little effect on the yield, although it could affect seed size and weight of the beans and milo. Del was glad the dry spell hadn't come a month earlier.

Del and Zach were in their farm shop before seven, repairing a rear wheel assembly on a brush hog, an implement that attached to the rear of a small tractor. It was already hot. The shop doors were flung open wide to receive any lingering breath of cool air. The tractor and brush hog were backed into the doorway.

"There's grease in the rear-wheel bearings," Zach said. Perspiration was already showing on his forehead. "Help me push this wheel assembly back in place?"

"I'm pushin', but the axle ain't movin.'"

"Push harder."

"Okay. I got it!"

"Put a bolt in the hole at the end of the axle, instead of a coder key. Those coders catch on brush and get pulled out."

"Got a washer for the end?"

"Here's a quarter-inch one. If it don't fit, we can probably make it fit. Drill it out if need be."

The little brush hog had two heavy, free-swinging knife-like blades attached beneath its flat, steel-hardened exterior, and was brawny enough to mulch saplings in a single pass, or anything else unfortunate enough to fall in front of its six-foot girth. It came with factory-installed wheels on each side, a model designed to be towed behind a tractor rather than mounted onto a tractor's three-point hitch, but that was when tractors were smaller. Today most brush hogs were triple the size of this one. Del and Zach kept this one around for small jobs, trimming around corrals and buildings. To make it more useful, they removed the side wheels, welded a three-point hitch mount to the front, and fastened a small, free-rotating tail-wheel behind. In an afternoon it was transformed into a tractor-mounted implement easier to maneuver in tight places and close to fences.

Harlan arrived about eight thirty, late, as had been usual for the last few months. Since earlier in the summer, he'd lived alternately with and apart from his wife, but reconciliations inevitably led to arguments and separation. At the moment they were apart.

"Hi, Harlan. How you doing today?" Del said, hoping his greeting sounded enthusiastic.

"Oh, I'd be doin' real good if I could get up in the mornings. I can't hear my alarm down there," he croaked, his voice hoarse and weak. For almost a week, Harlan had been sleeping in the basement of a fishing buddy's house.

"Man, it was like a sauna down there when I woke up."

"Harlan, you ought to make a clean break from that woman," Del advised. "I've told you that before."

"Don't want to throw away seventeen years," Harlan responded.

"Seventeen years of what? Nothing but problems. You used to enjoy yourself. Fishing on weekends. Now you hardly even do that."

"Yeah, I know. I need to get my boat fixed. A damn tree limb fell on it a couple weeks back. Broke the windshield. Knocked out the rear window in my truck too. Now I got two busted windows to fix. Actually, I got three. Last spring the glass on the driver's side door got shattered."

For a while the side window was a crinkled mosaic of fractured glass. Harlan ignored it. Then it fell out, leaving glass shards and laminate on the seat and floor. Since then he'd been closing the door over a black plastic bag to block the window hole if the weather looked bad. He wedged a piece of cardboard in the broken rear window.

Harlan knew it was a "gone fishin'" kind of truck. He didn't care as long as he could keep it running. A new one would have been nice, but he derived a certain perverse pleasure in telling friends how much work he'd done to this one. The cab was rusty but original. Almost everything else was a replacement—engine, transmission, most of the suspension, and the pickup bed, which was black and didn't match the maroon cab or a gray door. Legions of dings scarred the exterior. Two bullet holes in the bed were the result of Harlan's failed efforts to align the scope on his deer rifle. He'd braced the rifle across one side of the truck bed to fire at a target on the opposite side. The scope was further out of alignment than he thought. The first heavy bullet ripped through the top edge of the truck bed on the opposite side. Figuring he missed the target, he fired again, scoring another direct hit on the edge of the bed before he realized what he'd done. He hammered the bullet-torn grooves flat so the metal wouldn't snag anything.

Del changed the subject. "Harlan, you need to finish hauling those big bales off the Hyatt prairie. Use the Farmall 1086. If you finish before lunch, start brush hogging around the edges of the fields. We just got it fixed and we'll take it over there later this morning for you. Weeds are higher than my head along the fencerows."

"You want me to start mowing pastures too?"

"Not yet. We'll get the big brush hog ready for pasture mowing. But with that and manure hauling, you'll be busy until corn harvest. You got a lot of pasture mowing to do too."

"Don't I know it," Harlan responded, shaking his head. "Lots of rocks to watch out for in some of those pastures."

Even during summer months, Del and Zach worked in the farm machine shop when they had free time. Late the evening before, with slightly cooler temperatures, Zach tested his new pulling tractor in a pasture behind the shop, the culmination of nine months of work. On the outside it almost resembled a normal farm tractor, but the engine would have been more at home on a dragstrip. It was more powerful and more temperamental than anything he'd worked on previously. When he spotted a leaking seal in the transmission and another in the radiator, he also realized he had a lot more work to do.

Caitlin carried trash to a burn barrel behind the shop. On her way back to the house, she stopped at the shop with Crissy at her heels. The dog paused momentarily at her feet to pant, then hustled around the shop, tail wagging perilously close to electrical power equipment, portable lights, and containers of used motor oil. Seeing the dog, Boy the shop cat scrambled to safety atop the shop bench, arched, and then made a leaping getaway when the dog was occupied elsewhere.

"I've never seen you spooked like that before," Caitlin said. "You've been around tractors and pulling since you were little. I guess you've never handled anything so powerful."

Zach looked chagrined and Del laughed. "He hasn't even done that much to the engine yet. Wait 'til he finishes. It'll really smoke then."

Later that morning, Zach and Del took two tractors and a hay baler to storage in a rental farm shed a few miles away. Del seldom left machinery outside in the weather, but finding a place to store everything was becoming a problem. He already had machinery squirreled away in barns on three farms besides his own. If he removed a combine, which he'd need soon, there would be space for the tractors and baler.

Zach stayed at the rental farm to bulldoze manure, an unenviable job that had to be done every year because of manure built up around the corrals and feeding areas during the winter. By now the manure was dry. Harlan would load it in a manure spreader and put it on pastures when he wasn't occupied with other work. Del's spreader wasn't much different than the ones he'd used years ago, just bigger. A series of slats hooked between two endless drag chains scraped along the bed of the spreader and dragged the manure to the rear where an auger and rotating cylinders with spikes chewed into the load, flinging chunks high and wide behind. The band of dung-spotted earth left in the spreader's wake was hard to miss.

Del disconnected the grain header from his 915 combine, left the header in the shed, and connected an eight-row corn header to the combine. He had planned to have a newer corn header by now but hadn't been able to find one. That purchase would have to wait another season.

For corn harvesting, Del added two more rasp bars to the combine's threshing cylinder, reversed two pulleys to decrease cylinder speed to eliminate cracking the kernels, widened the cylinder threshing gap to accommodate the cobs passing through the machine, and increased the fan speed and air flow. Getting it right was part science, part trial and error. Del figured the best way to make sure it was right was to examine the grain in the grain bin and then walk a hundred-yard walk behind a combine. That could tell you plenty. You didn't have to find much grain on the ground to be losing several bushels an acre, and that was like throwing money on the ground.

Older combines, like Del's 915, which dated to the 1970s, required making adjustments externally, but after the mid-1990s most adjustments could be made remotely. Del often thought about the convenience of a newer combine, but it was a hundred- to a hundred-and-fifty-thousand-dollar cost he wasn't willing to make, at least not yet. Brand new ones could go for nearly four hundred thousand dollars.

Del drove the combine back to the homeplace and parked it in front of the machine shop. Pausing, he made a list of a few more parts and supplies he'd need to get it ready for corn harvest. He fished his cellphone out of his pocket to check the time. He could make a trip to Ashton and get back in time for a late noon meal.

A mile west of the farm, the blacktop road curved around a house and barn built on a low hill. Then the road ran arrow-straight for ten miles, rising and falling a little with the contour of the land. It was a familiar landscape, one he'd known since childhood. He knew every square-cornered field of soybeans and corn, the green pastures with black or brown cattle, the little streams hidden by galleries of trees, a T-junction where a blacktop road ran to the south, and a dozen unpaved roads that led off at perpendicular angles from the highway. In the distance a plume of dust rose from a vehicle on a dusty road. He passed several white frame houses with large front porches.

As Del drove he went over, in his head, the items he needed to purchase—oil filters, oil, transmission fluid, grease cartridges, a V-belt he didn't have, and several gaskets, among other things. He had it written down, but going over it, he thought he might remember something he'd forgotten. He'd also stop at A&C's Tractor Repair for the replacement emblems for an antique tractor Zach had restored. Del and Zach took pride in maintaining their farm equipment, and they also lavished attention on old tractor restoration projects, even down to small details like decals and emblems. A&C was on the way out of town, a convenient last stop.

Del often bought parts from A&C's and called their mechanics for advice. The shop was a large frame building with a sheet-metal exterior. Without air conditioning, the building had become unbearably hot during the recent heatwave, and the mechanics had pulled a couple of sections of metal siding off the front of the building to increase ventilation. Bare studs and pink insulation protruded around the gap. Del took the shortest route inside, stepping through the gaping hole in the wall and ducking slightly to avoid a diagonal 2 x 4 support before heading for the parts counter in a separate room on the left. On the way, he greeted a couple of the mechanics. Five well-worn tractors in the shop were in the midst of teardowns. The place was a chaos of transmission innards and metal parts lying exposed beneath the glare of overhead shop lights. The engine of one tractor was open with cylinders and pistons exposed. The front and rear halves of two others were separated, the halves suspended by chains and hoists. Both machines straddled pans of sludgy black oil strategically placed to collect drips from transmissions. Tool caddies, replacement parts, nuts, bolts, power cords, compressor hoses, chains, and blocks added to the clutter. Two industrial-sized floor fans in the middle of the shop roared furiously, replacing hot inside air with more hot air from outside. Del glanced at the fans. At least the air was moving. More tractors outside in a sunbaked parking lot waited like ailing patients for their turn in the shop.

Greeting Leroy, the parts manager, Del said, "Zach told me you had those little Farmall emblems he ordered. Said they'd come it."

"Don't think they've come in yet," Leroy grumbled, hardly looking up.

He wiped his calloused hand across his forehead where lines of sweat had formed. Del ignored Leroy's gruffness. It was just his usual manner.

"Hey, dingbat!" Leroy yelled at an assistant. "Can you check on this for me? Them metal emblems."

Without a word, a gaunt teenager with wisps of peach fuzz on his chin, a gold earring, and an overly large Floor Layer's Local 512 T-shirt sauntered up to the counter and grinned at Del.

"Check the part number for those emblems on the computer?"

"Aw, ain't never learned to do that on the computer," the boy drawled, almost as if his ignorance of the machine was a badge of honor.

"Goddamn it, then look it up in the parts book," Leroy yelled from the back of a long parts alley. "Look up the part numbers for those International Harvester emblems. See if we got them in. I don't think we did."

Standing in front of a massive twelve-inch-thick parts manual mounted on the desk, the boy thumbed through grease-smudged pages for several minutes, turning pages slowly, running a finger down endless columns of parts descriptions and tiny numbers. Then he checked an inventory list. Finally, he looked up and, with an air of finality, announced, "Nope, we ain't got it."

"You sure you looked up the right number?" Leroy yelled from the back of the parts department.

"Yep. Number ought to be 70-7086."

"Well, then we did get 'em."

"No, we ain't. Skips from 83 to 89."

"Oh. Told you Zach didn't know what he was talking about."

"Can you reorder them?"

"Yup. Come a dozen to a packet."

"How much are they each? Zach don't need a dozen, and I'm sure he don't wanna pay three or four dollars each for them."

"Naw, only a buck seventy-nine each."

"Okay, order a packet. He might even want all of them. Call us when they come in."

Del turned and walked back through the shop, leaving through the same gaping hole in the side of the building where he'd entered. He felt the hot breeze from one of the fans as he left.

A few miles east of Ashton, Del turned off the blacktop highway and drove a mile south to look at his milo crop on some rented land. It was drier here than on the homeplace even though only eleven miles separated the farms. A week ago this milo had immature seed heads, short and barely protruding above the leaves. Today the stalks were longer, but seed maturity lagged fields near his homeplace. He wouldn't worry yet. Milo did better in dry weather than corn or soybeans.

At their midday meal, Harlan looked tired and was unusually quiet. He often lingered to relax a bit after eating, and sometimes he even fell asleep on the couch in front of the television for a few minutes. Today he hardly ate anything, excused himself, and went back to work with scarcely a word.

"I hope he gets himself straightened out soon," Del said to Zach. "He's been like that most of the summer. Jack Larousse saw him in the field a few days ago, tractor stopped, Harlan outside pacing around, his arms waving. Jack thought he was having mechanical trouble and stopped to help. Wasn't nothin' wrong. Just Harlan on his cellphone, yelling at his wife. Jack said Harlan was all flushed and sweaty. Wasn't sure he ought to leave him alone to operate heavy machinery."

Zach shook his head. "He told me he had a two-hundred-dollar phone bill last month. I don't know how it could be that much. He don't have that kind of money to spend. He'll lose his phone if he's not careful."

Zach returned to the shop after dinner. There was a pull the next night, and he discovered he needed to install an automatic shutoff on the diesel engine. He figured he could get it in Prairie Point.

Del vacuum-tested the air conditioner on his cab-over tractor-trailer. He used a pump from an old refrigerator to pull a vacuum and clear the system of moisture. Satisfied there were no leaks, he charged the air conditioner. He used the truck mostly to haul grain to nearby elevators and didn't need air conditioning for short hauls, but a few weeks back he contracted to deliver five tractor-trailer loads of hay to a buyer in southern Missouri. Suddenly, fixing the truck's air conditioner seemed like a good idea, an irony not lost on Zach, who teased him about his newfound interest in air conditioning.

Zach left mid-afternoon to get the shut-off device from a parts store. He said his eyes were burning from the diesel fumes in the shop and he needed to get away for a while. A few days earlier, when he was chipping at a hot weld, a piece of slag had struck him in the eye, burning a tiny spot over the iris. It hurt so much he'd gone to the emergency room at the hospital where Amber worked. The doctor flushed his eye but said there wasn't any permanent damage. Del wondered if Zach's eye was still bothering him.

By the time Zach left, a rumble of clouds cast slow-moving shadows across fields, and a breeze, like a cool whisper, brought a bit of relief to oven-hot fields. Rain almost seemed possible. A family of goldfinches in gaudy black and yellow dress lit on thistles at the edge of Caitlin's garden. They swayed in the breeze and fluttered and pecked at the tiny seeds. On a roadside fence a skittish, loose-knit group of kingbirds, already heeding flocking urges, wheeled and sallied aerobatically for flying insects. Del looked at the sky and contemplated closing the shop doors, then decided it would blow over.

Later in the afternoon, Del threw a half dozen sacks of Big Red Cattle Supplement onto the back of his black feeding truck, figuring he had time to distribute some of the supplement to his cowherds. Harlan returned before he left, driving the little Farmall H with the brush hog they'd repaired earlier in the day. Del knew Harlan would be mowing inside the corrals, so he quickly penned up the four calves that had been given the run of the corrals by the barn.

As he put the calves in the barn, Del thought about what had brought these four together. They were growing fast and were much bigger now, but still playful with tails wagging and heels kicking. They bounded into the small wooden holding pen in the barn almost as soon as he opened the door. Del suspected their enthusiasm was prompted by lingering memories of being fed there in the spring. Last winter, as newborns, they'd all lost their mothers. Caitlin had patiently hand-fed each of them in this little corral, first with a prepared bottle formula, later teaching them to drink from a bucket. Two were twins whose mothers didn't have enough milk for two. The mother of another had fallen into the icy pond; a day after she was rescued, she'd given birth, just before she died. The mother of the fourth had a dry udder from a mastitis infection and had to be sold. Joined for a season, the four orphans would remain together until fall roundup and calf sales. Then, along with other calves born late last winter, their histories would become intermingled and lost. After the fall sale, they would be merged into increasingly homogenous groups and funneled into large commercial feedlots. Del, like all farmers and ranchers, enjoyed his animals and took pride in caring for them because he knew that if he didn't, they wouldn't take care of him. But he was pragmatic about their lives and fates. The four calves would bring a good price, as good as any others when they were sold, and it was money he and Caitlin wouldn't otherwise have realized. It was the same each season, only sometimes there were more.

After looking at the cattle herds, Del spotted Zach and Amber's gold and white spaniel, Maggie. She bounded out of a ditch not far from a neighbor's house. Caitlin had found her as a pup along a highway where someone had dumped her. She had given the pup to Zach and Amber, and Maggie willingly took up residence with them. Then, with Zach and Amber gone most days, and Nathaniel with a babysitter, Maggie grew up and began to wander. She took up with a neighbor. The neighbor tried to give her back. Maggie would stay a night, then leave. That had happened several times earlier in the summer. Eventually, Amber told the neighbor to keep the dog if he wanted her.

Zach, in an unflappable manner, shrugged and said later, "Better your dog running off than your wife."

27

September 12
Waiting to Harvest

September. During this busy month, farmers harvest corn, grain sorghum, and early-planted soybeans; grain dryers scream round-the-clock to dry down stored grain; pasture mowing and manure hauling continues; auction companies begin making short video clips of calves for late fall sales; black and yellow garden spiders and webs common around buildings; monarch butterflies migrate south; sumac shows first fall color

Del was ready to harvest corn. He had tested a sample for moisture at the co-op a few days before. At seventeen percent, it was still too high to sell without incurring a dock even though the dented kernels were hard and flinty. Worried about the weather, he decided to start the harvest anyway, figuring he'd store it in his grain bins and dry it using the grain dryers. Then it rained a little over Labor Day weekend, and again a few days later. Now the fields weren't dry enough to support the weight of combines and trucks. Del and Zach liked to start corn harvest by Labor Day. That gave them time to get the milo, a smaller harvest, immediately afterwards, and then start soybeans. He looked at the two big red combines parked inside his machine shed. Both had been ready for some time, their platform grain headers replaced with pointed, corn-picking heads. Caitlin noticed his anxiety. She could tell it in his conversations, which were mostly about the weather, field conditions, crops, and moisture content of the seeds. If fields stayed wet for a while, there was a risk of losing part of the crop. Until the fields dried there was little he could do.

For decades, corn prices had been chronically low, but the use of corn for ethanol production changed that, gobbling nearly forty percent of the annual corn harvest in recent years and creating demand that boosted prices dramatically. This year Del and Zach had planted over six hundred acres. For decades, American farmers had grown too much corn, but they did it because of government price supports—farming the government as much as the land, they'd say. Even with low support prices, growing corn was profitable if you grew a lot of it efficiently, and American corn farmers had become extremely efficient. Thanks to increasingly sophisticated varieties of hybrid seed corn, complex fertilizer schemes that matched soil types, better herbicides, and quicker methods of applying them, modern farmers had become the most efficient food producers in the history of humankind. And it didn't hurt that soils and rainfall were ideal for corn in the Upper Midwest. Del figured about the only way to get land like that was to marry into it or inherit it.

While they waited for harvest conditions to improve, two problems surfaced. One morning Del found a dead cow in one herd, then a dead calf in a second herd. He was on the phone with the vet even before he left the pastures. The vet's diagnosis was sobering—the cow likely had anaplasmosis, a disease that destroyed red blood cells. Their blood turned watery, their eyelids pale pink. A parasite transmitted by a horsefly bite was the culprit. The vet said to boost the amount of tetracycline in a feed supplement to prevent it spreading. Calves and younger cattle usually survived it, but older ones were vulnerable without treatment.

Then came worse news. The vet thought the calf had blackleg, a disease that struck fear in cattle owners. He'd confirm it when he stopped by later in the day. It affected mainly young calves and was almost always fatal. Calves with blackleg got gangrenous infections, especially in their legs, and died quickly. It could wipe out a herd of calves. Del had vaccinated for blackleg during spring roundup—the usual seven-way vaccine that was supposed to be effective against several kinds of *Clostridial* bacteria—but sometimes the vaccine only lasted a few months. The vet said maybe recent rains floated spores to the soil surface where they infected the calf, but it was difficult to know the cause for sure. All the calves in that herd had to be revaccinated immediately. That meant a roundup. Del hadn't planned on working any cattle so early in the fall, especially during harvest, but this couldn't wait.

The vet said he'd do the vaccinations and also pregnancy checks on the cows in that herd at the same time. It would take a day to get the cattle penned up and cows and calves separated. They'd revaccinate the following day.

Pregnancy checks turned up a few cows and heifers that weren't pregnant. That wasn't surprising. If they hadn't bred, Del never kept them another year. Some heifers might breed when older, but the expense of carrying an open heifer for a second year seldom paid off. He'd sell them and use the money to buy bred heifers or cows.

Three days later, with clearing skies and warm temperatures, the corn harvest looked like a go, but before they started Del backed his truck and stock trailer up to the white corral by the barn on the homeplace. Inside the corral were three Black Angus heifers and four older black cows, non-breeders culled from the vet's pregnancy tests.

"This is the last truck ride for you'uns," Del said as Zach moved around behind the milling animals, calling to them in a low voice and herding them, with outstretched arms and whip in hand, toward the narrow chute.

The heifers loaded easily, moving straight into the long trailer, but a cow, perhaps recalling a previous bad experience, doubled back. Despite Zach's sharp whip, she charged straight past him and continued to the back of the corral. Encouraged by the rebellion, the others followed. Zach jumped aside. "Nothing you can do but get out of the way," he yelled to Del, then made his way to the back of the corral again.

Del ducked inside the trailer, grabbed an interior gate, and slammed it shut, confining the three heifers inside to the front half of the trailer before they could escape, then he almost reflexively leaped backward as a splatter of thin manure from a frightened animal squirted through the gate.

With the men yelling encouragement, the four errant cows then ran straight up into the trailer. Zach quickly closed the rear trailer gate and they breathed a sigh of relief. "That wasn't too difficult," Zach said, catching his breath.

"No, but I had a close call inside there," Del said, grinning and pointing to a wet manure streak running the length of his pant leg. "One of them heifers had pretty good aim."

"Her last revenge," Zach said, laughing as he watched Del wipe the worst of it off with a handful of weeds from the edge of the corral.

"I'll take these gals to the sale barn. You work on the foundations for the new steel bins until I get back. Then we'll try the corn on the Somerset place. Ought to be dry by midday. By the way, has Harlan showed up for work yet?"

"He tried to call you earlier. I guess you didn't have your cellphone on. Said he was sick or something and couldn't get up."

Del grinned. "Yeah, that's more normal for him. Last week he set a record. Came to work every day, and almost on time. Hasn't done that in a while."

"Did he make up with his wife or something?"

"For about a week. But they got in a fight again over the weekend. Yesterday he told me Judd was about to kick him out of the basement where he sleeps when he's on the outs with his wife. Didn't know what he was going to do. Said he guessed he'd have to sleep in his storage unit for a few days. Got most of his stuff there already."

A line of trucks and trailers had already formed at the auction yard by the time Del arrived. Cattle were being brought in for the Monday night sale. Unloading was rapid. Truckers pulled forward next to a cattle chute. Over rattling diesel engines and bawling cattle, men in cowboy hats yelled to drivers where to stop and then swung heavy metal gates out to enclose the backs of trailers so cattle could be funneled into a long runway. While noisy trailer doors were opening and men were shouting to move the cattle, other men, including one of the auctioneers, counted the animals held briefly in the long corral alleyway and affixed numbered identification stickers to the backs of each animal. A woman in boots and western jeans stood beside the chute and recorded owners' names and addresses, prepared receipts, and handed copies to sellers. Barely two minute later, they waved Del forward to make room for the next truck. He parked at the side of the big lot and walked back to ask the women if the on-duty vet could do another pregnancy check on his cows. She nodded and handed him another receipt. "It's worth it," she said. "Don't cost much to check 'em again. No sense in selling one that's pregnant."

"That'd be like throwing money away. I do enough of that farming."

Someone alongside the corral overhead Del and said, "Buying and selling these old cows ain't all that different than farming neither. Mighty easy to lose money. You heard about the farmer that won the lottery, didn't you?"

Del shook his head.

"Yeah, somebody asked him what he was going to do with all that money, and he said he was gonna keep farming 'til it was all gone."

Del grinned. "That'd be funny except for being true sometimes." By the time Del picked up his receipt, another load of cattle had been unloaded and processed. It would be like that most of the day.

At home Del backed the stock trailer into the long shed at the south side of the machine shop, then climbed into the cab of his blue semi. The truck cab had a sour smell. He wrinkled his nose as he started the engine.

Zach saw him and grinned. "How's that dead mouse doing?"

"Wish I could get rid of it, but it's hard to find them. Follow me over to the Somerset field and pick me up. Then we'll move the combines over. I think it's dry enough to start pickin' corn." Corn harvest was underway.

28

September 18
Picking Corn

Modern corn hybrids are capable of remarkable yields. In the Upper Midwest, two hundred fifty bushels to the acre yields are possible under favorable conditions. Statewide average yields for Iowa and Illinois regularly exceed 170 bushels per acre; several other states are not far behind. Corn has one of the most pedigreed genetic backgrounds of any plant grown by humans, but for all its genetic manipulation it is vulnerable to drought. Newer flex-corn hybrids carry genetic coding for drought resistance, but early summer droughts can still decrease yields by fifty percent or more. It's every corn farmer's worst nightmare.

It was Del's worry too. That was why spring planting always started early. Rainfall is most predictable in spring and early summer, and watching the drought unfold in August only reinforced Del's conviction that planting corn as early as possible was important. In fairy tales about industrious ants and dilatory grasshoppers, there were always rewards for hard work. It was true in farming, too, but sometimes even unflagging effort wasn't enough. Still, it didn't hurt to try and handicap the odds.

From lowly grass in Mexico to genetically modified, super-sized monoculture in Iowa, *Zea mays* had come a long way, genetically. What we know as corn today—in a midwestern cornfield, a supermarket, or anywhere else in the world—never existed in a wild state. Scholars think it was domesticated in central Mexico six or seven thousand years ago from an inauspicious little grass called *teosinte*. None of the varieties of teosinte resemble modern corn, and they produce less than a dozen seeds lined up in a single row on a tiny "cob" not even two inches long. Yet for all their differences, teosinte and modern corn can still crossbreed, and occasional hybrids turn up in the wild. Teosinte, however, is barely a footnote at the beginning of a long history of human-assisted evolution leading to the biological and commercial goliath we know today as corn.

Columbus took corn from the West Indies to Spain, and from Spain and Portugal it was carried around the world. In 2003 farmers worldwide grew almost twenty-four billion bushels of it. That was the first time in history that corn production exceeded that of wheat. By 2009, worldwide corn production still surpassed wheat, as well as rice. It now seems poised to permanently overtake wheat as the world's most abundant commodity. The United States grows forty percent of the world's corn. Iowa, Illinois, and Nebraska lead the way. A half dozen other states are close behind. Worldwide, the United States is followed by China, Brazil, Mexico, Argentina, India, and France, but all of it, no matter where it grows, originated in warm, semitropical Mexico. Now it's grown in climates as dry as deserts, albeit with generous irrigation, and as wet as rainforests. Iowa and the Midwest cornbelt region suit it perfectly.

Modern corn is a fecund, seed-producing marvel, a grass that converts sunlight to food more prolifically than any other domesticated crop. From hybrids with growing seasons, maturity dates, and soil preferences as finely tuned as a racing engine, corn plants reach heights of six to nine feet, produce two dozen or more wavy, sharp-edged leaves, and send roots down four feet deep. Male flowers are produced in tassels at the top of the plant. The most visible part of the female flower is the silk, which extends from the tip of the embryonic ear located halfway down the stalk. Midway through the summer growing season, tassels produce millions of tiny pollen grains that scatter on air currents, pollinating the same plant and nearby plants, or drifting with the wind into other fields. When a pollen grain touches a strand of silk, it extends a tiny filament through the tubular silk to its attachment point with an embryonic ovary. From the five-hundred to fifteen-hundred fertilized ovaries on an ear of corn, a roughly equal number of kernels will develop in up to eighteen rows.

Kernels of corn are so tightly packed on the cob, and the cob so tightly bound by the husk, that modern corn cannot germinate or sprout without humans, or their machines, to first shell the crowded kernels. It is, in essence, a product of human engineering and would cease to exist without its human benefactors who free each kernel. The collective fates of corn and humans are now so intertwined that they are inseparable—corn cannot exist for long without people nor, it seems, can people do without corn.

The varieties of corn seem almost limitless. They differ in color, in size, and in ability to thrive around the world. Almost three hundred varieties of corn existed before Columbus reached the Americas, among them blue, yellow, and red corn, sweet corn, dent corn, flint corn, flour corn, hominy corn, and popcorn. Modern plant geneticists have crossbred some of these varieties to produce vigorous hybrids, and farmers worldwide have watched yields soar in recent decades. Of this diversity, however, it is mainly hybrids of yellow dent corn that provide the basis for our nation's corn food industry. Ironically, dent corn is the least recognized variety outside of agricultural circles. Most people know it simply as "field corn." The corn that Del and Zach, and all of their neighbors were harvesting was a hybrid yellow dent corn, so named because the flattened top end of each kernel has a small depression or "dent" in it. On the other hand, the sweet corn we see in supermarkets always has rounded kernels.

Although many varieties of corn exist, modern agriculture depends heavily on only a handful of genetic varieties, and most corn grown in the United States today is derived from about sixteen varieties. This worries geneticists. Extreme dependence on so few genetic lines is risky. Without the genetic diversity derived from many varieties, a species many lack the flexibility to survive physical and environmental changes that are inevitable over time. Losing the rich heritage of genetic variability in corn is an issue that has stimulated interest in preserving vintage seeds.

Del and Zach's concerns this morning, however, were mostly about saving their crop. They usually started harvesting corn about ten or eleven in the morning, when the stalks were crisp and dry and the husks rattled, but it was damp and foggy this morning. Harvests always presented farmers with a conundrum—wait for a crop to dry down to safe levels and then rain could make fields impassable, or wind could blow down a crop. On the other hand, harvest too early and the combines are apt to waste corn because shelling corn from damp, tough husks is more difficult. There also is the issue of storing grain with high moisture because it can overheat, become moldy, even spontaneously combust if not dried down to acceptable levels. Above a certain moisture limit, a farmer would always get below market price, if he could even sell it.

Del and Zach, like most farmers, were practiced weather watchers, and they didn't like gambling with their crops. If that meant harvesting when moisture levels were running a little high, they'd do it to get crops out of the field. They could always dry it down later, but running a bunch of electricity-hogging grain driers day and night for six weeks could add hundreds of dollars to an electric bill. It was one more expense that squeezed already razor-thin profit margins. Right now Del figured it wouldn't hurt to wait a few hours before starting the day's harvest.

Early morning light filtered through the fog like orange gossamer spread over the gently rolling farmland. For a while the countryside existed in soft focus. In the dampness Del pulled on a light-colored canvas jacket with a dark corduroy collar and stuck a pair of yellow work gloves in his hip pocket as he left the porch. He figured the haze would burn off soon enough. He could use the morning hours to unload the three trucks from the previous evening's harvest.

The grain elevator at the co-op was slow, a small-capacity facility dating from an era when people brought grain to sell in farm wagons and pickups and small trucks, not the eighteen-wheelers farmers used today. Time had eclipsed many smaller grain elevator operations like the one in Prairie Point. Once, these tall metal buildings topped with spidery, stilt-like auger pipes were a fixture in upper midwestern towns, but now there were fewer of them. Across the Great Plains, too, where wheat was king, a similar change was taking place, although grain elevators there were usually larger—rows of giant concrete cylinders linked together like monstrous honeycombs. Visible for miles, the North American wheat belt landscape was literally defined by these grain elevators. Now many were closing their doors too. Where local co-ops once served hundreds of area farmers with small acreages, they might now serve only a few dozen with much larger acreages. Farmers also were increasingly bypassing these facilities, trucking harvests overnight directly to international grain buyers in places like Kansas City where prices were better and unloading facilities faster. But trucking to more distant facilities involved time and expense, and Del thought it was a wash. Convenience ought to count for something. Besides, using nearby co-ops supported local communities, and those people were his neighbors.

After weighing his load of corn, Del pulled the truck up onto the unloading ramp at the co-op and set the air brakes, which hissed loudly as the expelled air blew a cloud of dust up beneath the truck. Del climbed down from the high cab. Walking back along the side of the long, white trailer, he rotated a large metal wheel to open the trailer's belly dump. A thick band of orange-yellow corn rattled downward, spewing onto the elevator's Archimedes' screw auger set in a narrow concrete trough beneath the unloading ramp. In seconds, the stream of corn smothered the auger and formed an elongated pyramid that dissolved in on itself even as it was fed from above by grain pouring out of the truck. A fog of corn dust crept out beneath the dumpsite and soon engulfed the truck and the unloading ramp in an itchy haze. Dwayne, an elevator employee, his cap and untrimmed beard mealy with corn dust, stayed to push some of the burgeoning pile of overflow corn back toward the ramp intake with a broad scoop shovel. He took little notice of the dust. Del stepped away from the expanding orb and then walked across to the co-op office on the opposite side of the weight scales.

Crossing the scales, he glanced upward and noticed a flock of white pelicans milling overhead, white pinwheels against a cyanine sky. They were almost directly above the galvanized metal feeder augers radiating from the apex of the elevator. It was odd to see them here, circling over town. Without a single wing flap, they were gaining altitude, wheeling upward with each sweeping rotation, already perhaps five hundred feet up. As he paused to watch, they slid silently off the apex of the invisible thermal and began drifting southward, effortlessly aligning themselves into a lopsided V. He wondered where they were going. They could be in Arkansas by nightfall. As they disappeared, he stepped inside the co-op office to chat with the supervisor while his truck unloaded.

By the time he returned home, warm sunshine bathed the countryside. Golden fields of corn were dry now, cornhusks rustling like sandpaper in the gentle autumn breeze. With luck he'd get an hour of harvesting before lunch.

Del started the combine engine and engaged the thrashing mechanism. Letting the engine idle, he climbed back down and gave the machine a last-minute walk-around check for problems. He noticed a loose header chain but decided to ignore it, climbed back up inside the cab, and steered the big red machine down the road almost a mile and into the field. Zach had picked most of the corn in the higher, well-drained part of the field, so it took a few minutes to get to the bottomland.

Zach was at the homeplace working on foundations for two large steel grain bins they'd bought from a retiring neighbor. Once the foundations were ready, their plan was to put metal skids under the cylindrical bins and slide them across some fields to the homeplace and line them up next to the bins already in place. Zach was on his knees with a level and measuring tape when his phone rang.

"What's wrong?" he asked, hearing his dad's voice.

"Need you to bring the black truck and tools. The 1680's got a busted hydraulic line. Same as what happened to the 915 last June."

"Wasn't planning on finishing this field today anyway," Del said with black humor as Zach arrived with the truck and tools.

"I see you broke a chain too?" Zach said, noticing that Del was reconnecting the links in the chain on the combine header.

"Chain broke was why I stopped. That's when I saw oil all over the side of the combine. Lucky to make it up here to the edge of the field before all the oil drained out. Once the oil's gone, this thing ain't goin' nowhere. I seen that chain was a little loose this morning when I greased the combine, but I thought it would hold."

"Tightener's already pushed up to the top of the slot," Zach said.

"Fixing the chain's easy. I already took the broken link out. I'll drain the rest of the oil into a bucket. You need to get under the combine and loosen the coupling on the oil line that leads to the hydrostat motor."

"Looks like it's been rubbing against a bracket that holds another oil line," Zach noted as he bent down to get under the combine. On his back in weeds and dirt beneath the combine, Zach put a leg against the long wrench, but the coupler wouldn't budge. "Where's the pipe wrench?"

"In the truck, I think, if Harlan ain't moved it."

"Hand it to me. I'll try it. Handle's longer."

"Here, try this hammer on the pipe wrench handle. Maybe you can jar it loose."

Zach strained against the pipe wrench, then pounded on it with the heavy hammer. Chaff and grease from the underbelly of the combine dropped into his hair and onto his face with each blow. In the cramped space he grimaced, banging hard on the wrench handle. Eventually the coupling loosened, and the oil remaining in the line began to leak out, some of it dripping on his already grease-smudged hands. He wiped some of it on his black t-shirt.

Watching him, Del laughed. "Leastways the oil won't show so bad on that t-shirt." Leaning into the narrow space behind the huge left front tire,

Del held a second wrench on the coupler from above to keep the two parts of the coupler from slipping, but there was so little room to maneuver he couldn't move the wrench.

"This thing's a pain to get loose," Zach said. "Just like the one on the other combine. Engineers never think about making things easy."

"They don't care. It's not them that has to fix things."

It took the better part of an hour to unscrew the coupler and remove the long section of steel oil line that snaked down forward along the side of the combine before looping back 180 degrees to join the hydrostat pump behind the front axle. Once freed, Del examined the tiny hole worn on the metal oil line, rubbing a calloused finger across the nearly invisible hole. "I'll take this up to Hayden's at Belle River," Del said. "I just called him and he's there. He's not preaching or marrying anybody today. He can braze it."

"We could try to weld it here."

"We'd probably just burn a bigger hole in it. Brazing will work better. Why don't you go eat dinner while I'm gone? I'll eat later."

By mid-afternoon, they had the broken oil line fixed. Zach wiped an accumulation of grease from his hands with pumice-laced hand cleaner and a dry towel. He studied the newly installed oil line for a minute, then climbed up to the cab and started the combine engine. Amidst a clatter of noise and a fog of dust, he steered it back into the field. Corn harvest was underway again. It was mid-afternoon when Del finally ate. Wendy and her baby daughter, now almost eight months old, were at the house. Del stayed around to talk and to hold his granddaughter. With a full-time job and new baby, Wendy hadn't been able to visit often. Del noticed her husband wasn't along. "Jonathan working?" he asked.

"He's on day shifts and couldn't come. You know he hates to miss a trip up here."

Later Del drove his old blue Chevy semi into Prairie Point to get it weighed at the co-op. He'd need the empty weight to calculate the landowner's share of corn, which he would sell, but he was planning to store his part, hoping the price would rise later.

By the time Del got back to the field, Zach had filled one truck, and the combine tank was also nearly full. As the big red combine approached, Del could see yellow corn heaped above the combine's flared tank extensions like a giant, bulging egg yolk. Zach raised the header at the end of the rows and taxied toward the side of the field, meeting the truck halfway. As

the combine approached, the whine of the engine increased to a crescendo. Dust swirled over the truck, then cleared somewhat as the combine unloaded a thick stream of corn into the trailer. Minutes later, with the combine noise fading, Del stood alone beside the truck. He looked across the field and then at the ground, littered with dry, shredded leaves. The rows of sharp-cut cornstalk stubs severed close to the ground stretched away in lines like stitches in a gigantic needlework. Walking slowly between a couple of rows, he looked for kernels of corn, a cob that wasn't completely shelled, anything indicating the combine wasn't adjusted properly. He knew that some guys would hog through a field, push too hard, and leave grain in order to finish quickly. There were times it couldn't be helped, but he didn't like it. He wanted to do a good job.

The field looked clean. Walking back to his truck, he noticed a sprinkle of orange spots, like tiny winking lights fluttering and drifting across the field. Monarchs! Their headway was erratic and unhurried, almost aimless, but there was no mistaking their intent. He'd seen a few butterflies moving southward earlier, not big numbers, just a few trickling across the countryside, one here, one or two there, in a broad, silent front over fields and pastures. They came every September, but today there were more. With blind instinct they pressed southward to a winter rendezvous with others of their kind in Mexico. The monarchs, like the fields they were flying over, signaled closure to the largess of summer. Del relaxed a little. It felt good to have the corn harvest underway.

29

September 24
Last Tractor Pull

Beneath a lens of gray rainclouds, the highway looked wet and shiny. Seams of black tar crisscrossed blotches of light and dark pavement like a crazy quilt in monochrome. Zach stared ahead, ignoring the flickering pavement that laced the road. Spotting his landmark, an old rusty combine in a field to the right, he slowed the truck and turned onto a narrow gravel road. It was a rough and uneven strip of proliferating potholes. A crude, hand-painted sign with a red up arrow pointed ahead. A mile beyond another red arrow pointed to the right where a narrow dirt track crossed a soybean field. On the far side of the field, the track led to a small opening in a thick row of Osage orange trees. Zach thought the location looked like a clandestine site for something illegal, but there was no doubting the place. A line of trucks and cars behind him and several ahead were all making their way slowly across the soybean field.

At the far end of the field, just before the trees, a heavyset man in bib overalls and baseball cap, and a woman in jeans, sat on a pickup truck tailgate. They looked up at Zach's tractors, chained to his trailer, and with little expression waved him ahead without the fee that spectators would be charged. Behind the trees the track ended, dissolving into a grassy field that had been turned into a parking area where cars, pickups, SUVs and camper vehicles were lining up. In the middle of the field, a hundred-yard-long graded dirt track, ruler flat and wider than a two-lane highway, had been gouged from fresh earth, a brown scar in a green pasture. A rise of metal bleachers with bare wooden seats flanked the right side of the track, and on the left, two long flatbed trailers with wooden flooring were positioned parallel to it. A couple rows of metal folding chairs atop the trailers provided additional seating perilously close to trackside action.

Zach swung his pickup to the left and parked at the far end of an area designated for pullers. A long row of pickups, a dozen eighteen-wheelers, and an assortment of other tow rigs were already parked there, including flatbeds, enclosed trailers, and one with a sleeping unit mounted on the front of the trailer. Brightly painted tractors and stiff-framed, four-wheel-drive pulling trucks were being unloaded, the trucks mostly along a back fence, the tractors nearer the weigh-in scales. Men and women hovered around trucks and tractors. A few wore jackets with colorful logos proclaiming the names of their pulling outfits. One group had a canopied tent and folding chairs set up near their truck.

This was the final competition of the state tractor pullers' association, the culmination of a summer-long season of weekend pulls that began in June. It was a sport that involved a commitment of time and money and was repaid far more in love than purse winnings. It was also often a family affair. With men and women frequently traveling considerable distances to events, the support of their families was important. Every one of them lived for the "full pull," hoping their tractor or truck, in a rush of smoke and noise, would yank the weighted sled the full three-hundred-foot limit—the puller's holy grail. Those who didn't share their passion for the noise and horsepower might think it was madness.

There was a tingle of excitement and expectation in the air as tractors and pulling trucks were unloaded. Some of the pullers were clustered in groups, swapping stories with friends, or relaxing in aluminum folding chairs. Others were registering or lining up for weigh-in. A deafening roar momentarily filled the air as someone started an engine, revved it to insanely high RPMs, and then shut it down.

Del and Zach began removing chains and straps from Zach's two tractors while Caitlin and Amber unloaded folding chairs, and a baby blanket and toys for Nathaniel. Zach was still in the hunt for point standing, but his mind was already on next season and his new tractor, the still unpainted and unproven machine that had been his dream since last winter when he started building it. He'd pulled it for the first time a week ago. Tonight would be his last chance to check for problems at an officially sanctioned event.

The meet began promptly at five in the evening when a young girl in tight jeans and cowboy boots stood on the flatbed trailer and sang the national anthem. The gathering crowd doffed caps and hats and stood while recorded music broadcast over a scratchy public address system provided accompaniment. Cars and trucks continued to arrive well after the anthem and opening announcements. Vehicles were packed with families—husbands, wives, children, grandparents, and grandchildren. Later in the evening, carloads of teens would show up.

Tractor pulling, as a sport, shares some of the insanity of drag racing although the goals of the two differ. A dragster seeks the quickest quarter mile, a puller the longest distance irrespective of time. Pulling is more like drag racing with a heavy sled towed behind, one whose coefficient of friction increases almost exponentially with distance. The two sports do share one important denominator—both are expensive. Equipment, performance parts, safety compliance, and travel converge on bank accounts of puller and dragster aficionado alike, and both sports often evolve as extensions of businesses. Pullers, however, usually come from families in agriculture or trucking and rarely have sponsors. Drag racing's popularity and media coverage, on the other hand, attracts deep-pocketed corporate sponsors.

Zach drew the first pull of the evening, facing an array of tractors with quixotic names like Wild Delight, Orange Blossom Special, Gold Digger, Basket Case, and one, like his own, with no name. With his tractor's rear tires squat and digging and its front end barely touching the track, his pull of 282 feet 10 inches bested his nearest competitor by just over eight feet. It wasn't a full pull, but it was close, and the crowd roared their approval. Afterwards, he rejoined Del and Caitlin and Amber and Nathaniel, who had taken up positions near the 200-foot marker on the track. Zach was pleased as he stood beside Del, the two of them reviewing the pull. He noted that the organizers weren't using spikes on the weighted sled tonight, something they sometimes did to make the sled more difficult to pull.

"I been trying to get the sled operators to stop dropping the spikes for pulls in my performance class. Tonight they didn't, and it sure helped me."

"Hard track helped too," Del added.

Zach's success with the pull, however, hardly seemed to matter to him. He was already thinking about the pull later tonight, the one with his dream tractor. He was so excited he could hardly wait.

After dark the temperature dropped. Dampness crept in, and sweaters and jackets were brought out from trucks and cars. The calm evening air held a redolent blend of exhaust fumes, engine heat, greasy food, and fresh mowed grass. Portable food concessions did brisk businesses in hamburgers, hot dogs, and funnel cakes, and lines soon formed at several blue Johnny-on-the-Spots behind the bleachers. The announcer droned names, hometowns, and statistics of pullers over a public address system that was frequently overwhelmed by the howls and screams of powerful machines. In one specialty category called prostock, the machines were long and low, with razor-cut lugged tires, and looked more like real dragsters than tractors. Some had two, even three screaming engines mounted in a line up front.

"I'd like to get away from these things for a while," Del said. "They're awful noisy." Caitlin and Amber and Nathaniel had already moved away from the sidelines, taking their folding chairs with them. Del and Zach followed and decided to wander through the back lot where the tractors were parked. Owners enjoyed talking about their tractors, but it was conversation few but mechanics and pulling enthusiasts would have found comprehensible, mostly about engine parts, transmissions, turbos, gear ratios, and weight reductions.

"This one doesn't have any brackets mounting the wheels to the axle," Zach noted. "He's welded the axle solid to the wheel. Probably saved two hundred pounds."

Del spotted a diesel with a huge turbocharger and a water line leading to the intake. "Here's a guy who's injecting water into the cylinders through the turbo."

Zach nodded. "A lot of pullers want their fuel cold. They say it makes it more explosive. A couple of the tractors I pulled against tonight run fuel lines through dry ice."

"Cool enough tonight anyway," Del added. "Diesels ought to be pulling good."

The two of them walked back trackside. A photographer stood poised outside the chalked, out-of-bounds line of the track, snapping pictures of each pull. As Del and Zach looked on, a dual-engine prostock entry caught fire during its pull. There was an explosion, and a fireball erupted from the side of the lead engine immediately opposite the photographer. He dived for the sidelines, cradling his cameras as he went down. Then he looked around, realized he was unhurt, and scrambled back to his feet and began snapping pictures as a flagman and track assistants rushed to spray fire retardant on the flames. The puller and tractor were enveloped in an expanding miasma of white retardant, steam, and smoke. Track assistants disappeared and reappeared in the swirling melee as they continued spewing retardant. The driver, in helmet and fire-retardant suit, remained strapped inside the tractor's steel safety cage in stoic resignation. He was still seated as his broken machine was towed from the track.

Del shook his head. "This being the last pull of the season, things are starting to break. Been a long summer."

"This next guy will probably blow something too," Zach said, looking at the starting gate where a tractor was hooking to the sled. "This guy never knows when to let up on an engine. He goes flat out until something breaks."

"I thought that's pretty much what all of them did," Dale countered.

"Well, that's more or less true, but this guy's crazy."

It was a prescient remark. His engine blew, just as Zach suspected, and so did another of the following three entries; engine fires, stripped gears, blown hoses—all pushed beyond mechanical limits. None of the entries in this class used radiators. They ran the engines hot, calculating they would last for the duration of the pull without mechanical cooling. Pulls usually lasted only twelve to fifteen seconds, but the strain heated metals and alloys to the limit. In seconds, temperatures soared. Manifolds belched plumes of fire and glowed cherry red at the end of fiery pulls.

Breakdowns weren't confined to high-end prostock and superstock classes. A Kool-Aid–colored orange Allis in a lower weight category blew a radiator, sending a spray of steam and boiling water backward onto the driver, a gritty, determined type who, despite the scalding water, never flinched, bulldogging ahead until the tractor spun out, defeated by the weight of the sled. Sitting immobile in a bath of steam rising from the engine compartment, the unflappable driver, with helmet and goggles still in place, seemed untroubled by his brush with disaster. Several onlookers shook their heads.

Someone wisecracked, "He wants them points pretty bad, don't he? I would've stopped when that damn radiator blew."

"He likes scalding water in his face better than I do," another remarked dryly.

By midway through the evening, a knot of teens had gathered on one of the flatbed trailers overlooking the track, and an overflow loitered behind or on the trailer steps, milling, posturing, always talking. They hardly noticed the noise of the pulls, which periodically approached thresholds of pain. Fewer still even looked trackside as they fiddled nervously with cigarettes, opened their eyes a little too wide, or laughed too loud. Some clung to each other, arms draped in studied casualness over a companion's shoulder. It was a semi-autonomous fraternization of mostly rural youth, and from it emerged a kind of collective energy that seemed to take on a life of its own, at once a part of the affairs of the evening and also separate from them. From time to time, teens drifted away, wandering among long shadows cast by floodlights on rows of tractors and trucks. Then they would stroll back again.

One threesome threaded their way through a noisy group of small boys playing an impromptu game of tag football with a miniature ball. "Go long," a boy yelled, and bodies and football were swallowed in darkness beyond a parked row of trucks and tractors. For a moment the two groups merged, then separated, each moving on, as oblivious to the other as stars in intersecting galaxies.

Beneath the bright lights of the track, another ear-splitting surge of noise and display of horsepower built to a crescendo. The crowd cheered, but on the dimmed sidelines the focus of youth took scant notice. Standing beside Amber, Zach put a hand on her shoulder, squeezed gently, and nodded toward the crowd of teens around the trailer. She knew what he was thinking. "We were there once, weren't we?" she said, looking up and smiling as she balanced Nathaniel on her knees.

While waiting for his next pull, which Zach knew wouldn't be until nearly midnight, he and Del wandered back through the dozens of tractors parked behind the bleachers. One fellow was working on his tractor engine in semi-darkness behind the track. His father, a bent figure in bib overalls and cap, held a small flashlight with failing batteries. Tiny bolts, nuts, washers, sections of fuel line, and a fuel pump were laid out on a piece of sheet metal beside the tractor. A large toolbox was on the ground beside a front wheel.

Upset with the performance of his engine, the fellow complained, "Going to have to tear this thing down next winter. It ain't been running right for the last three pulls. Too dark to make any changes tonight, I guess."

Del nodded and decided the fellow was too upset to talk. Walking on, Del and Zach greeted several men they knew standing beside a black pickup truck with a wheelchair in the back and a shiny red tractor chained to a tow trailer behind. The men were listening to the fellow in the truck. "You heard about Jake?" the fellow asked, leaning out the window.

The men nodded and shuffled uncomfortably.

"What do you reckon will happen to him now after his accident?" somebody asked. "I heard he lost part of a leg and the other was broke in six places. Lost his thumb, too, and part of a hand."

"Yeah. Got caught up in a sweep auger inside a grain bin he was cleaning out."

"A rough deal if you ask me," another said.

"Damn lucky there was a couple guys nearby that heard him hollerin'. That thing woulda killed him if they hadn't got it shut off."

"He'll be back pulling again next year, I bet," Del said, joining the group. "I heard he was going to reposition some controls on his tractor. Maybe add a hand brake so he can use his left hand."

The men nodded in mute agreement, and there was a momentary silence. Finally, one said, "Yeah, nothing will keep him from pulling again. He's as crazy as us."

Farming was often cited as one of the most dangerous occupations, although most farmers probably wouldn't regard their occupation as particularly dangerous. But there were risks working around large animals, operating heavy machinery, transporting farm equipment on public roads, and lifting heavy items. Nearly all farmers knew somebody who had gotten injured or killed in a farming accident—a tractor overturning, a careless finger or hand or piece of clothing caught in a pulley or chain or spinning power shaft. Early farm equipment was particularly dangerous, generally lacking even the most basic safety features. Del thought that urban dwellers that took up farming later in life also skewed statistics, being more likely to suffer injuries than those who grew up farming.

Del and Zach left the group and began backtracking around the far end of the track to find the photographer. "Most of these guys would never give up farming or a hobby like pulling even if they did get hurt," Del said. "Just like that guy back there with his wheelchair."

"He was in a motorcycle accident, wasn't he?"

"Yeah, almost twenty years ago. Hasn't walked since, but he loves his tractor and these meets. Spends his weekdays working on it. A friend drives it for him. 'My Lady in Red,' he calls that tractor. Got the words painted right there over the engine, and I guess she is."

Zach wanted to see if the photographer had any pictures of one of his pulls earlier in the summer. "Pictures are all in little packets in that bag over by the fence," the photographer said, yelling over the noise as he bolted back trackside to snap photos of a four-wheel-drive pickup screaming up the track in a cacophony of noise and spinning tires spewing dirt.

Seconds later he was back, breathless, his two cameras, one digital, one film, dangling from his neck. "These pictures here are from the Fort Scott pull, and these are from the McLouth pull," he said, handing Zach a couple of thick folders of photos. "Looking for one in particular?"

"I'll just look through the last three or four meets," Zach said, kneeling down away from the shadows to get better light on the pictures. There were hundreds of photos from pulls all over a multi-state area, all neatly labeled by date and location.

"Here, take this light," the photographer said, handing him a small but unusually bright light. "It's a video camera light. Probably the most expensive flashlight you'll ever use. Damn thing don't work on my cameras. Makes a great flashlight, though."

Zach kneeled down to examine the pictures and finally picked out three. When the photographer returned, he handed the guy a ten. "Sorry, I don't have exact change."

"Me neither. Here, take this five and the pictures. One of those pictures ain't that good anyway."

Sitting on his tractor, waiting for his final pull, Zach absentmindedly watched swarms of insects orbiting the floodlights alongside the track, their bodies spinning in frantic circles like electrons around a nucleus. It was almost midnight. Fog was beginning to descend. In the heavy, damp air, the insects seemed transfixed, as if trapped, whirling endlessly in gossamer orbits. By morning the fog would be much denser. Zach wanted to get home before the highways got too foggy, but he was determined to see how the tractor handled, what gear worked best, where to position front and rear weights. Everything happened frighteningly fast. Practice runs were essential.

Nine months ago, this tractor had been his dream, a transmission and differential sitting on frozen ground outside their farm shop, an engine on blocks at the back of the shop. Now the engine screamed. Amidst a plume of black exhaust and wheel-spinning dirt, he gripped the steering wheel hard. The tractor's front wheels were off the ground for most of the pull. He breathed a sigh of relief that he'd been able to control it. At the end of the track, he listened as they called his name and pull distance over the loudspeaker. Third place. It wasn't the worst showing. Now he had a better idea what he needed to do before next season.

Within minutes of the last pull, now past midnight, the bustling parking lot began to empty. Families, friends, spectators, and pullers with their bright machines in tow streamed across the pasture and out to the main road in a line of conical light beams rising and falling in the darkness with each bump in the road. In an hour the pasture would be empty and quiet. The smoky air would clear. Fog would silently settle over the rolling countryside. For the rest of the year cattle would graze in this pasture, and birds would search for insects and seeds in the grass, and crickets would trill on warm summer nights. And then, in another year, these men and women would return, perhaps on a similar perfect fall evening, and the drama of noise and smoke and power would be played out again.

30

September 27
Milo

The two red International Harvester combines lumbered back and forth across the field like gigantic Pac-Men in larger-than-life drama. With engines screaming and thin lines of diesel smoke spewing rearward, they gobbled broad sweeps of milo with methodical efficiency while enormous front tires two and a half feet wide and taller than a human stenciled thick herringbone patterns into the soil. As the cavernous grain tank behind the cab filled, Zach glanced up at a mirror to monitor its contents. Satisfied it could hold no more without spilling, he wheeled the combine away from the rows of grain and pulled alongside one of the two eighteen-wheeler grain trucks waiting at the side of the field. Even before he came to a full stop, he swung the long, tubular unloading auger out, locking it in place. When he engaged the auger, there was a thunderous rattling from the Archimedes' screw inside, and a telephone-pole-thick geyser of red milo began gushing into the truck. Milo dust enveloped the truck and combine and drifted away across the field. A few minutes later, the last of more than two hundred bushels of milo rattled out of the auger and onto the conical seed pile in the truck.

Rains had slowed fall harvests again. At the beginning of the month, Del and Zach were delayed several days getting the corn harvest underway. Then rains held up the start of milo harvest for a few days. Although they were finally underway several days ago, the fields were still soft and even the huge balloon tires on the combines made imprints in the soil. Del and Zach had to leave the trucks on the roads or in the grassy sod along the sides of the fields. Now fields were firmer, and they could drive the trucks closer to the combines to unload. The day before, Zach and Harlan had harvested over three thousand bushels of milo on one of the rental farms, about the same as on previous days. The yield was good—eighty to ninety bushels an acre.

"Rain in August would have pushed the yield even higher," Del told Caitlin.

"Well, you shouldn't complain. It's still a good harvest."

"Yeah, but it's always something. If the harvest is good, the price isn't. What you gain one place you'll usually give back somewhere else."

"Oh, you farmers are never satisfied," Caitlin teased.

Del's phone rang, interrupting their conversation. "First truck's loaded." It was Zach calling from the field to signal he and Harlan would begin unloading their combines into the second truck.

"See there," Caitlin continued. "You have more now than you can haul. You can hardly keep ahead of the combines."

Del grinned. She was right, but he hated to admit it. It was 1:30 in the afternoon. From now until dark, it would be a race to stay ahead of the combines and keep an empty truck in the field. By evening they would fill all three tractor-trailers twice.

Before he unrolled the tarp over the first load, Del picked up a handful of milo. It felt dry and hard like tiny BBs. This variety was dull reddish in color. Others were whitish or yellowish, even bronze. As he looked at the milo, an ironic smile spread across his face. If people in urban areas recognized it at all, it would likely be as one of the seeds in their birdfeed mix, but that was probably the least important market for it.

Milo had been domesticated in the Old World for as long as corn in the New World, but it wasn't as useful. It also had many confusing names—kaffir corn, durra, shattercane, grain sorghum, and just plain sorghum among others—and even more names in Africa and Asia. Botanists called it *Sorghum bicolor*, but even here there was apt to be disagreement owing to its many varieties. There were sweet sorghums grown to make molasses, and grass sorghums for cattle forage, and broom sorghums for fibers. The one Del planted was a grain sorghum grown for seed. The most notorious sorghum was Johnson grass, an invasive from Turkey. Like so many well-meaning but misguided introductions, it proved almost impossible to eradicate. Named for the man that introduced it, Johnson grass grew to eight feet in height and choked out native plants. It was a disaster in fields of milo and corn, an unwanted "volunteer," and any herbicide that killed it also killed the closely related corn and milo. Del, like many farmers, had spent his share of time pulling it by hand.

Milo had long, waxy leaves and a bulked-up root system the envy of any modern corn hybrid, and it thrived in water-starved regions. Most of the seven- to eight-million acres of milo grown each year in North America were planted in the southern Great Plains from Nebraska to Texas, a region of miserly and capricious rainfall.

On average, farmers plant up to 75,000 milo seeds per acre and expect 85 percent of them to germinate, the pale green plants all crowded together in rows. Like corn, milo needs three to four months to reach maturity, but that's about the end of the similarity. The commonest variety of milo tops out at about four feet, a dwarf compared to corn, but easier to harvest because the seeds are produced at the top of the plant so there is far less plant material for combines to process and discard. Growers like milo because of its resistance to drought and diseases and pests, but its Achilles' heel was its appetite for nitrogen. It could extract a hundred pounds of nitrogen from an acre of soil and another fourteen pounds each of potassium and phosphorus. These were expensive to replace, so Del, like most farmers, rotated milo fields to soybeans the next year because of its nitrogen-fixing ability.

Milo seeds are mostly starch, but they are rich in antioxidants. It is almost tasteless but absorbs flavors well. In Africa and parts of Asia, it is eaten boiled, ground, or decorticated and used in porridge, unleavened bread, cookies, cakes, even malted beverages, but it has never caught on as a human food in the New World despite its gluten-free attributes.

By mid-afternoon, Zach and Harlan had harvested forty acres of milo across the highway from the homeplace, then moved a mile west and a mile and a half north to another rental farm. They'd try to get twenty acres there before dark, finish the rest on that farm the next morning, and move again by mid-afternoon. Del shuttled trucks between fields, storage bins, and the co-op, but by late afternoon the two combines were filling the trucks faster than he could unload them. Within an hour, Del was back at the co-op with his second load of the afternoon, but this time there was a truck ahead of him and he would have to wait for the guy to unload. Ducking inside the main building, he spotted the manager. "Hey, C.J.," he teased, "I think I'll take this load back home. Maybe bring it back here next winter when I've got a spare day or two to get a truck unloaded. Price might be better then, anyway."

"Yeah, or worse," C.J. said. "Suit yourself."

They both laughed. C.J. knew the unloading facilities were painfully slow, and that wasn't going to change, but grain prices almost certainly would change. "Problem with prices is that they're growing too much of this stuff in places like Brazil," C.J. said.

"Argentina, too, and Australia. Even Europe," Del added.

"I've read land's cheap down there in Brazil."

"So's labor."

"They're clearing millions of acres for farmland. A lot of it for the same crops we're growing here. Farmhands working for practically nothing. Hard to compete with that. We're all gonna be out of business if it continues."

Del figured there could be some truth to what C.J. was saying. It could happen, but it would take a while. The other thing was that commodity prices were good now, but they probably wouldn't stay that way. They were notoriously cyclical. "Unless prices stay high, maybe we're all gonna have to start farming in Brazil or someplace," Del said.

"A few guys are doing it already. University's got seminars on it. I heard you could pay for land in just a few years. But getting the stuff to market's a problem."

"In pioneer days if you wanted land you just moved west. Now farmers are headed south. Have to learn to talk Portuguese or something if we do that," Del said.

On his way out of the co-op, he walked past a long row of farm and vet supplies and stopped at the old soft drink machine in the back corner. Next to it was a refrigerator with a large glass door in front. It was stocked with medicines for cattle and farm animals. He'd bought his share of cattle antibiotics and vaccines from that fridge, and he'd fed plenty of quarters into that soft drink machine too.

Del put four quarters in the machine, then opened the narrow glass door, reached inside, and tried to push a small tab-shaped lever down, which should have opened a gate releasing a can of soft drink, which would roll down an incline into a cylinder-shaped receptacle where he could pull it out. It wasn't the soft-drink industry's greatest engineering coup. The lever wouldn't budge. He rattled the door and pushed another Diet Coke lever. Nothing moved.

One of the secretaries at the front desk saw his predicament and came back with a key to open the machine. "It's old and doesn't work very well," she said.

"Well, that's for sure."

"Here, take what you want."

"Thanks," Del said with a laugh. "Thought I was going to have to pound that thing into scrap metal just to get a Coke, but it's already about scrap, isn't it?"

Harlan was running the 915 combine when a fanning mill belt broke. The big four-bladed, reel-like fan, located behind the front wheels and in the belly of the combine, provided a flow of air across the shaker shoes and sieves, which separated chaff from grain. Like a thousand other parts on a combine, a single belt or bearing malfunction usually disabled the entire combine. Replacing the belt wasn't complicated, but it would take a little time. Harlan decided to drive the machine the mile and a half back to the homeplace where he had access to more tools at the farm shop.

When Del returned from the co-op, he saw the combine in front of the machine shop. The ground beside the combine was already littered with oily rags, socket wrenches, and little piles of bolts and nuts that Harlan had removed. One heavy bracket remained, the only obstacle to freeing the big fan pulley so a new belt could be installed. While Harlan struggled with the bracket, Del pulled his loaded truck alongside the grain bins, opened the trailer's belly dump a few inches, and then started the Farmall H that powered the drive-over grain dump and auger.

The tractor coughed a little cloud of blue smoke and sputtered to life. The auger, like a giant Tinkertoy stretching up to the conical rooftop of a steel bin, clattered to life, and dust began to accumulate around the hopper's mouth, which shimmied and began gobbling the milo pouring from the drive-over dump beneath the truck bed. A tiny amount of grain leaked out under an elbow of the auger's hopper. Del ignored the leak, adjusted the trailer's belly dump so it wouldn't overflow the hopper, and went to look for a replacement fan belt for Harlan.

Del had spare belts in his shop, but none were the right size, so he checked in an old, metal-sided shed on a nearby rental farm. Stepping inside the old shed, with its slightly pungent smell of diesel and tires and old hay and oil-soaked earth, he looked up at the dimly lit back wall where he kept spare belts for emergencies like this. Dozens of belts, sorted by lengths and shapes, hung like dusty black snakes looped over pegs on the wall. Some were nearly new, others used. He scanned the wall, spotted two that looked like they were the right size, and took both. Back at the homeplace, he gave Harlan the belts, told him to use whichever one fit best, and took a grain truck back to the field where Zach was harvesting. By the time he arrived, Zach had filled both waiting trucks.

"There's enough time for me to get this last truck loaded before dark," Zach yelled amidst the noise and dust as he slammed the combine's cab door, revved the engine, and folded back the unloading auger.

The combine lumbered forward into the gathering dusk. For a minute, Del stood beside the truck, watching and listening as the big machine, in a cacophony of receding sound, faded from view like an apparition swallowed in dust.

Del grabbed the shiny handrail behind the truck door, swung up into the snub-nosed cab of the big tractor-trailer, and slid into the seat some five feet above ground. For such a large truck, the inside of the cab was cramped. He was glad he didn't have to spend a lot of time in it. The engine compartment took up most of the space in the middle of the cab, and as far as he was concerned, the sleeper compartment behind was just tool storage. With the engine roaring, he pulled out of the field and up onto a gravel road and turned left. A mile and a half later, topping a long low rise, he turned onto a blacktopped county road. Heading east toward Prairie Point and the co-op, he shifted without clutching through nine or ten of the fourteen forward gears.

On the way to the co-op to weigh the load, he decided to stop at the homeplace first and help Harlan finish replacing the fanning mill belt. By the time they finished, it was too late for Harlan to get back in the field. It didn't matter. They'd be ready tomorrow and would move both combines west to another rental farm with another hundred and twenty acres of milo. Two days' work barring breakdowns. Del remembered a breakdown a few years earlier, a rear combine axle that broke. It was the middle of fall harvest. Today they'd lost only a couple of hours fixing the belt. The axle had taken almost two days to repair, an eternity when crops were in the field.

This morning a weather report had caught Del's attention. A tropical storm in the Gulf of Mexico was passing over the western tip of Cuba and the Yucatan—a long way off, and it might not even reach the coast, but it had a name now and that, he figured, was a bad sign. If it moved northward it could, in a matter of days, affect somebody's harvest on a farm in the Midwest. That could affect the futures price of corn or milo or soybeans and, in turn, the price of cattle feed, eventually even the price of a steak on somebody's table. It was all connected.

Parking the truck on the long weight scales at the elevator, Del left his truck and went inside the co-op to get his weight ticket and have the clerk record his name and the landowner's names on the ticket. He attached the weight ticket to a little magnetic clip on the dash of the truck. It was a routine that never varied no matter how many loads of grain he took to the co-op. He'd store this load, like most others today, in the steel bins on the homeplace. Still, it all had to be weighed first.

At the end of the day, he'd unclip the weight tickets and take them to the house. First thing the next morning he'd record the weights of the grain and portions of ownership in his farm records book, and if he didn't have time Caitlin would do it. With so many properties and rental arrangements to look after, it was a lot of recordkeeping, and rental arrangements varied. In some agreements landowners got a third of the crop, and Del took two thirds for shouldering most of the production costs of seed, fertilizer, fuel, and labor. In a year when rains, weather, and prices aligned, both parties made a little money. Other landowners preferred cash rent—a flat fee—which guaranteed a predictable return no matter if the year was a bumper crop or a failure.

Caitlin heard him arriving and met him outside the house. It was almost dark. A few bright stars were just barely visible against a violet sky that turned fiery orange on the horizon. Del drove the white truck and trailer close to the yellow auger, swung the belly dump of the hopper under the trailer, and started the old red Farmall tractor again. It was a familiar routine. Caitlin had come out to tell him that his supper was ready. They stood for a moment, away from the dust and noise of the auger and tractor, and talked quietly in the gathering dusk. They could feel the air growing cool now, and a dampness made it feel cooler than it was.

"They say this winter is going to be a long, hard one," Caitlin said.

"Well, I'm having a long, hard time already this fall," Del said as he thought about the rain delays and breakdowns. "But the crops are turning out good. Who told you about the winter?"

"Harlan did. He said he cut open a persimmon the other day. It was full of spoons, you know, like the shape of something inside the fruit. He said when you see those, it's going to be a hard winter." Then she laughed. "I don't know why people say that. Maybe the spoons reminded somebody of shovels, and you need to shovel lots of snow during hard winters."

"Wonder where Harlan heard that?" Del replied, his voice trailing off as his thoughts turned back to the grain still unloading.

"Somebody also said that woolly worms had thicker coats this fall. That's supposed to mean a hard winter ahead too."

Del was gone now, already turning the shiny silver wheel that closed the belly dump of the trailer, and shutting down the tractor and auger. Abruptly everything went quiet. More stars were visible now, pinpoints of light in the darkening sky. There was an especially bright one low in the west. Caitlin thought it might be Venus. In the quiet of the evening, they walked together back toward the house, toward the familiar warm glow of lights in the kitchen and back porch windows. Tomorrow work would start all over again. Soybeans were almost ready.

31

October 4
Office with a View

October. Soybean harvest dominates work at beginning of month; grain driers dry stored grain; farmers attend Farmfest show; decisions are made about crop rotations for coming year; auction companies videotape calves for nationwide sales; cows pregnancy tested and nonbreeders sold; first freeze likely this month; cattle moved onto cornstalk ground for a few weeks; flocks of blue jays begin flying south; fall foliage color peaks by end of month.

By the beginning of October, Del said his soybeans were "made," and to any farmer that meant they were ready to harvest. With crops in the field and unpaid seed and fertilizer bills from spring planting filling mailboxes, midwestern farmers like Del were once again mobilizing combines and trucks for the season's last harvest. It was one they hoped would fill grain bins and granaries and keep bankers from paying visits. The morning was cool and damp. A veneer of gauze-thin clouds had skidded in on Kansas winds overnight. This morning the clouds revealed little of their intentions, but Del sensed a change in the weather. He didn't need change right now, so when the Montgomerys sat down to an early noon meal, there was an unspoken urgency. Caitlin sensed it. She hurriedly set out ham sandwiches, potato chips, apple cobbler, and ice tea, and Del, Zach, and Harlan gathered around the table. They ate quickly, foregoing a few minutes of relaxation in big overstuffed chairs in the living room after the noon meal.

Nathaniel, asleep on the floor in the living room, missed the meal. "He's been cranky most of the morning," Caitlin said. "It's either the weather or he's getting another tooth."

Zach looked in on Nathaniel but let him sleep. Like Del, he knew that during harvest nothing mattered but the crops. There would be time to spend with the family later. Del and Harlan were already making their way out the back porch and to their trucks.

A hundred acres of early-planted soybeans remained unharvested, more than a day's work for two combines. A little more than two hundred acres of late-planted soybeans would be ready in about two weeks. Del had checked the fields a few day before. The ground was firm enough to hold the weight of combines and trucks. He pulled a few soybean pods from a stalk, stripped the beans from their brittle pods, and bit into a few. They were slick and shiny and hard as bullets. Rain now could cost him part of this crop, maybe all of it if he didn't get it out. Ten days ago these fields were a monochrome of green, only a few leaves showing flecks of yellow the way soybeans did when they matured. Then, almost overnight, the leaves yellowed and withered. Now, with the leaves gone, the fields were dull and dry, a two-dimensional landscape of dry beanstalks, each carrying up to twenty small seedpods. The pods were fuzzy and lumpy and hung in pairs or whorls of three along the stem from near the ground to the tip of each two-and-a-half-foot tall plant. The regimental order imposed upon the crop by the planter last spring remained, but where the field was once vibrant and alive, bursting with life beneath the embrace of a summer sun, the plants now looked withered and exhausted, like lines of weeds after the first hard freeze.

Because of wet weather in the spring, Del had been able to plant only part of his soybean crop during the normal May period, and it was these that he was harvesting. The rest of the crop had been planted in early June, a few weeks later than he'd liked. These late-planted beans wouldn't be ready to harvest for about two more weeks, only a little ahead of double-cropped beans, which a few of his neighbors had no-till planted in late June after harvesting wheat. Late-planted and double-cropped soybeans were a gamble because summer rains were never a sure bet. If rains failed, so did the crop, and with it the loss of seed, fertilizer, herbicide, fuel, and a farmer's investment of time. Del, like many farmers, sometimes took the risk. With profit margins thin, double-cropping was a chance to squeeze two crops a year from a single field—another variable in an already uncertain financial stew.

Zach parked his pickup with toolbox and tools next to the two combines at the edge of the soybean field. Harlan left the blue Chevy semi at the side of the gravel road, and Del pulled his white eighteen-wheeler into the field behind Zach's service truck. Within minutes, both combines were rumbling southward through the long field of soybeans, each machine barely outpacing an opaque dust cloud that coalesced behind them. Del watched, relieved that soybean harvest was going well. The sound of the machines was music to his ears, an orchestra of well-oiled machines beneath a lupine sky.

Later in the afternoon, Ross, one of his landlords, stopped to watch. As Zach turned a corner, he spotted Ross's lanky frame standing at the side of the field. There wasn't time to stop and talk, but he waved to Ross to join him inside the combine cab.

With quick strides, Ross caught up to the machine as it slowed, and was then immediately enveloped in a swirling cloud of dust and chaff spewing from the combine's tail. He could feel the vibration of the big machine, even feel the ground quake underfoot as more than thirty thousand pounds of steel rumbled to a halt.

"Climb up into my office," Zach yelled to Ross over the noise and dust as he reached out with a brawny arm to swing open the glass-paneled entry door to the cab.

Making his way up the steel ladder and across the catwalk over the top of the big lugged front wheel, Ross slid into an auxiliary seat next to Zach, closed the heavy door, and was instantly transported into another world. It was, as Zach said, a mobile office, insulated from the dust and cacophonous noise outside by quarter-inch-thick panes of shatterproof, floor-to-ceiling glass windows and soundproof seals. Eight feet up and cantilevered over the business end of this red leviathan, the cab was almost whisper quiet, even as on the outside the engine screamed, the sickle clattered, and a mind-numbing array of moving parts and spinning belts rumbled in a chaos of dust. The combine, a late 1980s model 1680 Case-International, was the largest made at the time. Zach handled it as nimbly as a toy, even though it steered from its lugged rear wheels. Like a mobile harvesting factory set loose on an agricultural cornucopia of epic proportions, it was a machine built to match the vast acreage of America's heartland.

After the glass door slammed with a resounding thud and Ross and Zach's world went quiet, the machine moved forward, its big engine smoothly powering up to field speed. The air inside the spacious cab was filtered, and it could be heated or cooled to suit the operator. A radio mounted overhead broadcast news from a war zone halfway around the world. From the middle of a sun-dappled soybean field in October, its relevance seemed remote. The view from the cab's glass paneling was panoramic and stunningly close to the action. Everything happened immediately in front of and below the cab, almost at an operator's feet, and it happened with unnerving speed.

The combine's gaping twenty-five-foot maw and matching-length batting reel, which resembled nothing so much as an elongated, riverboat paddle wheel, gathered everything in the machine's path with uncanny efficiency. The batting reel assured that little escaped, pushing the crop, and even agile insects, backwards onto the combine's platform header. Dry beanstalks and their precious cargo of seeds were guided between conical-shaped sickle guards like hair in a comb. There they met the zigzag edges of the cutting knives attached to the sickle bar, which oscillated back and forth in a blur. Everything in the path of the sickle was amputated with guillotine-like efficiency. Even stray field rats and rabbits suffering a moment's indecision were at peril. From his seat inside the cab, Ross peered obliquely down through the glass at the maelstrom below, watched the machine stripping the field razor clean, and, for a moment, felt uneasiness in his stomach. He realized he was leaning too far forward, concentrating too much on the incessant inflow of the crop and the hypnotic rotation of the batting reel and auger. Yet, fascinated, he continued to watch, almost transfixed, as the revolving reel pushed the dry stalks and debris rearward where it was instantly grabbed by a large rotating auger that swirled everything inexorably toward the central throat of the combine. There, all matter, animate and otherwise, abruptly disappeared, as if consumed in a paroxysm, lost in an insatiable black hole. What happened thereafter, once the crop moved into the raging heart of the machine, happened out of sight, and was infinitely more mysterious if less graphic.

Zach explained that International combines were rotary types that employed a spinning cylinder, like a giant, elongated squirrel cage with heavy, rasp-like bars spiraling around the outside; it was positioned with its axis parallel to the inflow of grain. Other combines thrashed by passing the grain once beneath a cylinder that spun perpendicular to the inflow of grain. Both ultimately did the same thing, employing rasp bars on the cylinder to knock the grain loose. Then a powerful fan blew the lighter stems, leaves, and plant parts—the chaff—upward and rearward while the heavier grain fell through a series of sieves onto a catchment pan at the bottom of the machine where it was augured up into the grain tank. The straw and chaff, meanwhile, were walked rearward on pans with rear-projecting teeth that looked like old scrubbing boards. The pan walkers collected and saved any overflow grain while the straw and chaff exited the rear.

Modern combines are marvelously versatile machines, a fact not lost on Zach, who added that the miracle of a combine is that they are able to harvest the tiniest clover and grass seeds or shell heavy corn off foot-long ears, or anything in between. And it was all done in seconds while simultaneously handling a massive volume of dry leaves and stalks. In Missouri, an acre of fescue might produce only a few hundred pounds of seed, but corn yielding a hundred and seventy-five bushels an acre produces ten thousand pounds of seed, not counting the weight of dry plant material running through the machine. Such dramatic difference in seed weight and plant material moving through a combine require numerous internal adjustments so the grain can be cleaned, remain undamaged, and not be lost out the back of the combine. And now, as Zach noted, most of those adjustments could be made by an operator from the comfort of the combine cab.

Looking back over his head, Ross could see the grain tank and its overflow extensions. The tank alone held as much grain as an average-sized, single-axle truck. A thick stream of soybeans, invisible to the operators in the cab, was pouring into the grain tank, and the shrill din of noise it made outside was barely a muffled swishing sound inside the cab. From his seat, Ross felt as if he were inside a sterile enclosure, like a bubble boy isolated from the storms of life and noise and dust and pestilence that raged outside.

Zach saw Ross looking around. "Operators used to monitor their machines by listening to them. If a bearing was seizing up, or a belt slipping, or the threshing cylinder slowing, you could hear it."

Ross nodded, recalling that when he was a kid, his dad used to do that.

"But you had to sit out in the sun and insects and dust and mold all day. Lots of farmers suffered from the dust."

"City people think farmers are getting soft or lazy riding around in air-conditioned machinery," Ross added.

"Well, it's really a health issue. Unless you've experienced prickly dust and blistering heat all day, you wouldn't know. Course it's more comfortable.

But in a city office nobody's going to ask you to sit in a big cloud of dust and bugs all day, or in the burning sun. Farmers don't like to do that either."

"Yeah, but that's what we used to do," Ross said. "Nobody had enclosed cabs back then. Just little canvas umbrellas on our tractors in the summer. They were always getting torn up by tree limbs."

"And not much better in the winter."

"That's right. We had a canvas stretched between the tractor engine and fenders. That helped, but it was still cold. Farmers mostly used tractors and wagons to feed cattle then. Hardly anybody had four-wheel-drive pickups."

Because sealed cabs on tractors and combines insulated operators from machine sounds, at least subtle ones that could signal a mechanical problem, manufactures had to devise ways for operators to monitor their machinery. That led, ultimately, to the widespread use of electronic monitors. It was a major turning point in agriculture, and a necessity as farm equipment become more complex. Warning lights, monitors, and alarms became industry standard.

"This thing's got a warning light for practically everything. Here, watch," Zach said, cutting the engine speed slightly.

Immediately, an alarm sounded and several yellow lights flashed, warning that the threshing cylinder speed was too slow and that the engine RPM also was too slow.

Ross looked impressed. "Wow, that thing lit up like a fire alarm. Couldn't miss that."

"Beans got to be cut close to the ground too," Zach continued, pleased to have someone to talk to while he worked. "Soybeans have seedpods almost down to the ground. You're gonna miss a lot of beans if you don't cut practically at ground level."

Zach was harvesting at nearly six miles an hour, faster than a human could walk, and the cutter bar was slicing along mere inches above the ground, all the while instantly responding to the slightest irregularities in the field. The sorcery behind the combine's ability to avoid ploughing the giant header into the dirt, or into some obstacle in its path, lay in a series of electronic sensors along the bottom of the header, any one of which could instantly signal hydraulic cylinders to raise or lower the header platform. Older combines had wire sensors, a sort of a guide-by-wire system, but they didn't work very well. Before about 1970, combines didn't have any sensing mechanisms at all. It was up to the operator to watch and to raise and lower the header manually, a task impossible at harvest speeds today and with such wide header platforms.

As Zach and Harlan guided their combines steadily back and forth, the field shrank noticeably with each pass. The cloud of dust following each machine sometimes enveloped them completely when they harvested with a tailwind to their backs. Even if a combine was out of sight, the dust cloud could be seen. As the combines circled the field and reduced the remaining crop, panicked cotton rats began to flee their cover like the proverbial "rats leaving a sinking ship." Mostly they bounded along between two bean rows, trying to keep under cover, sometimes zigzagging ahead of the combines for dozens of yards before veering off at the last moment. Even if they outran the combines, their options were none too good. Ross noticed at least three red-tailed hawks watching from trees or fence posts at the edge of the field. For a while the hawks pounced on hapless rats at will, but eventually even these consummate hunters became satiated and lost interest. After that, a few rats successfully sprinted across several hundred yards of open terrain, in little marathons of terror, to safety in the grass at the edge of the field.

Occasionally, a tiny sparrow fluttered ahead of the combine, too, flying up low in erratic, zigzagging flight. They usually dived back into grassy cover beyond the field. Zach said he didn't know where they came from, but Ross thought they might be migrants from somewhere up north that had put down in the field to rest.

At one end of the field there was a small sinkhole. It was mostly filled now, but Zach swerved, maneuvering the combine around it. "That's all that's left of an old coalmine shaft," Zach said.

"So Peabody Coal didn't dig these?" Ross questioned.

"Oh, no. Peabody only dug strip pits. Whoever dug those didn't have big steam-powered shovels. Did it the hard way."

"By hand?"

"I think so. Seems incredible, but somebody had to dig down twenty or thirty feet to get a coal vein. Then worked out to the sides. See that bare area over there by the road?"

Ross nodded, squinting at an area without vegetation.

"We planted beans right through that, but not a single one came up. Probably too much coal dust in the soil. Ground's poisoned. That mine could have been dug more than a hundred years ago, but the land's still no good. We think that's where they brought up the coal and loaded it."

"Must have been dangerous in those little tunnels," Ross ventured, trying to visualize men crawling around in dark, poorly supported tunnels and picking at a thin vein of coal, filling a bucket or two at a time with nothing but a carbide lamp for light. "Musta dug it for their own use."

"I'm sure it was local. No OSHA around then, either," Zach said, laughing. "Most people around here now don't even know about these old mines. Things get forgotten. A generation or two and it's lost."

Slippery elm and hackberry and Osage orange brush now grew along one side of the old mineshaft, and it wasn't visible from the nearby road. Time had almost completely erased the lives and toils of the men and women living here just a few generations earlier. Eventually even these old mines would disappear.

By mid-afternoon, Del took the first of the day's harvest to town to weigh it. Minus the truck, it was about 38,000 pounds—over six hundred thirty bushels. He returned home and unloaded it. He hadn't hedged any of his grain this year because he didn't like the price of the calls. Maybe the price would increase later. It was a gamble.

At the homeplace, Caitlin gave Del sandwiches and water to take to Zach and Harlan. Del had the truck back in the field by five in the afternoon. Zach was grateful for the sandwich, but Harlan declined his, saying he had to leave in an hour. It was enough time for Del to unload another truckload of soybeans before taking over the combine after Harlan left.

With the combines running smoothly, they finished a series of mostly odd-shaped little bottomland fields sometime after seven in the evening. The yield—around thirty-five bushels an acre—was fair considering the drought in August. With a fifty-acre field remaining, they decided to move the combines this evening so everything would be in place for tomorrow.

It was a quiet gravel road, and they were just underway when the hydrostat pump on the 1680 blew a gasket. Del, driving the older machine in the rear, could see what had happened. He grabbed his phone. "Hey, Zach, you got an oil leak!"

Sitting up top beside the engine, the hydrostat pumped oil down to the front axle, providing a kind of gearless, variable-speed transmission for both forward and backward motion. Because of the high pressure, they tended to fail as they got older.

Standing atop the combine with two metal access panels open, Del mulled over what to do. "We've had some trouble with busted hydrostat lines this year, but this is the first gasket leak."

After a moment he said, "Zach, why don't you go back to your place and get the army truck? We can use it to pull this thing over to the next field. Hardly anybody can get by it now unless they drive down in the ditch."

"How am I going to get there?"

"Well, walk, I guess. It ain't that far. Less than a mile, I reckon. Trucks are still in the field, loaded with beans, and I don't have a pulling chain in either one of them anyway."

Zach looked sheepish, then turned and left on foot. While Del waited in the gathering dusk, a small pickup truck coasted quietly up behind the rear combine and stopped. It was Win, a neighbor's son, who had moved back into the community a few years ago. He had some flat sandrocks in the back of his truck.

"Hey, Del," he called. "What's goin' on?"

"I'm blocking the road," Del deadpanned.

"Yeah, looks that way."

"One of the combines just blew a hydrostat gasket. What you doing out so late?"

"Got the night off."

"You still working with the sheriff's department?"

"Yeah. Still undercover."

"How does that work? Everybody around here knows who you are."

"No, not around here. I'm working a few counties away. Well, I was until a couple days ago. A guy recognized me. That's why I'm off tonight."

"So, what happened?"

"This guy saw me. Said, 'Hey, do I know you?' And I said, 'Nope, I doubt it,' trying to look confident. 'Never seen you before in my life.' But then I got outta there quick as I could."

"Guess he knew you, huh?"

"Oh, yeah. When I was in uniform, I'd probably arrested him a half dozen times. Probably high on drugs most of the times. Now they'll be moving my surveillance area."

Del shook his head. "I'd hate to do what you're doing. I'd always be afraid of running into somebody like that."

"Sure you guys don't need any help? If not, I've got to get home. Wife's got me working on a landscaping project, so I been digging up some rocks over in my dad's field."

Before Win could maneuver his truck around the combine, Zach returned with the army truck, an old deuce and a half he bought a few years ago and fixed up. He drove the truck through the ditch, then up onto the road, and backed up in front of the disabled combine. Del wrapped a chain around the combine's front axle and fastened it to the truck hitch.

"Take it slow," Del yelled. "No power steering now, so this thing will be hard to steer."

The truck moved forward with a throaty purr. The chain jerked slightly, then tightened. All ten truck wheels bit the gravel. The combine inched forward slowly, its giant wheels crunching the gravel, and then it picked up a little speed.

Win maneuvered around the remaining combine and followed. When Del and Zach had the combine off the road and into the field for the night, he stopped. "What you got in that thing?" he yelled from the road. "No army truck ever sounded like that."

"We put a Chevy 350 V-8 in it," Zach replied with a sly grin. "Old engine didn't have much power. We bought two of these trucks; the other's for parts."

"Where'd you get 'em? They're hard to find."

"This one's been in the county for a long time. It's a 1953. City of Fairmont bought it for their fire department but never used it. Then a guy near Ashton bought it, but his kids got in trouble for driving across people's property. We got it from him."

"What about the other one?"

"Liquidation sale from a surplus dealer up north. It wasn't in good shape." Zach jumped out of the cab and ran around to the front of the truck. "Check these little red lights. They still work! Used for night conveys. Just so somebody close in front could see you in a rearview mirror."

Win bent down. He could see a tiny triangle of dim red light, like an LED keychain light, over each headlight.

"Gauges in the cab got the same little red lights," Zach added. "World War Two technology."

"Who woulda guessed they'd thought of stuff like that?"

As Win's truck slowly pulled away, his lights illuminated the trees and fence posts along the roadside in ghostly silhouette. Del and Zach drove back to get the other combine, which was still blocking the road, and moved it into the field alongside the first. Del figured it would be hard to find a hydrostat gasket, and he'd already decided to make one the next morning. He had some gasket material in his shop. Besides, even with only one combine working, they could finish the last field of soybeans tomorrow.

Finally, Del and Zach drove the two loaded tractor-trailers to the homeplace and backed them, side by side, into the long shed on the south side of the shop. It was dark now, only the big yard light and the lights inside the house intruding on the darkness. Del glanced up and saw a sky full of stars. It was what every farmer wanted to see at harvest. Still, as Zach started to leave for home, Del began climbing up the ladder to the top of the big circular grain bin to close the lid at the top where the elevator spout poked through.

"I'm no weatherman, but I don't want to have to get up in the middle of the night and come out here half naked and close a bunch of grain bins or move trucks," he said over the growl of Zach's truck engine.

"Me neither," Zach replied as he left the driveway.

Sagittarius leaned to the south now, poised to disappear from view. There was a light on in the kitchen window. Caitlin had supper waiting. It had been ready for hours.

32

October 8
Farmfest

Farm Talk, a weekly newspaper that serves agricultural communities across several midwestern states, said it best. "This weekend agriculture rules." The event was a three-day extravaganza of exhibits spread across several hundred acres of rolling, partly wooded hills north of Springfield—a tradeshow and wishing well of farm products designed to stimulate those "I sure could use that" urges in anyone with a few drops of country blood in their veins. Whether visitors were owners of farmettes, mega-farms, or merely curious onlookers, they would find, among the more than seven hundred and fifty exhibitors, an impressive sample of the diversity of livestock, machinery, trucks, agricultural accessories, and tools available.

"Too much to see in a day, really," Del said as he and Caitlin walked past yet another line of the latest farm machinery. Zach, with Amber pushing Nathaniel in a stroller, followed behind.

Amber said the machinery exhibits were overwhelming. "I don't know what hardly any of this stuff is," she said with a laugh, "so it gets a little boring after a while."

"Zach and I want to talk to some of these machinery reps," Del said to Caitlin. "I need a new cattle-loading chute, and Zach wants to look at the welding exhibits. If you and Amber want to go inside the pavilion with Nathaniel and look around, we can meet over in the tractor section in about two hours and eat our sandwiches."

"We'll stay with you for a while. Until we get bored, anyway," Caitlin said, grinning at Del. "We'd like to look for a toy tractor for Nathaniel, but that shouldn't take long."

It was a warm October day with a pale blue sky and a languid, hypnotic feeling in the air that lulled everyone into thinking summer could last an eternity. Fall colors were still almost a month away, but there were, even now, hints of change in the air. A few blue jays straggled overhead, harbingers of a mostly silent, winged migration already underway. In neglected corners and on banks and behind fences where mowers couldn't reach, the cheerful blues and yellows and golds of fugitive asters and sunflowers and goldenrods provided refuges of color from the manicured green of a landscape overburdened with lawnmowers. These flowers were reproductive procrastinators, the last native plants of the season to flower. Not one of them sent up so much as a single bloom until summer was nearly spent, but these hardy perennials, with pedigrees that stretched back ten thousand years or more, could easily survive an autumn freeze or two without compromising their reproductive futures.

A soft breeze, not hot like summer, but with the pungent scent of autumn leaves, tugged at sturdy oaks anchoring Ozark hillsides and, if the breezes hinted of a winter season ahead, nobody noticed. The monochromes of winter may have been just around the corner, but on this jubilee afternoon, they weren't on anyone's mind.

In a parking lot chockful of pickup trucks, license plate registrations revealed a demographic dispersion heavily tilted toward Missouri, Arkansas, Kansas, and Oklahoma in approximately that order. To a historian it might have seemed a political juxtaposition of the north and south, but to the men and women who mingled here today, it was just rural Midwest with a sprinkle of curious urban dwellers, all strolling past an extraordinary smorgasbord of agricultural offerings. Some came to look and perhaps to enjoy a day away from rural isolation and routine. Others came to make comparisons of products and prices. A few, like Del, planned to make purchases if the price was right. Food and drink vendors were much in evidence, event organizers having made sure there were enough hamburgers and hot dogs and funnelcakes and cold drinks on hand to energize even the most lethargic visitors. And for the foot weary there were benches and shady trees and demonstrations of products throughout the day.

It was an irony, perhaps, that this celebration of modern agricultural technology was attended by many Amish families, a group famous for their dedication to older, pre-combustion-engine methods of farming. Today, while these Amish families made little forays into a future filled with color and electronics and gasoline motors, their horses and buggies outside the fairgrounds waited patiently for their return to the present. Unmindful of this apparent dichotomy, Amish men and boys in wide-brimmed hats and denim overalls eagerly inspected brightly-painted tractors and machinery they would never be permitted to own or use, while inside the pavilions Amish women and girls in long dresses and bonnets studied up-to-date household and garden products. But for all of their wide-ranging curiosity, it was the livestock and horses that drew their most admiring comments and held their attention longest. They were, among other things, excellent purveyors of fine horses and good animal breeding stock. Leaving the livestock pavilion, Caitlin caught sight of one young Amish girl in bonnet and long dress, skipping along behind her mother. On her feet were fashionable Nike track shoes in what seemed a wayward blend of modern and traditional values. Or perhaps, even among Amish, an occasional allowance was made for youthful indiscretion.

Tradeshows like this one, and there were others relatively nearby—one in Kansas City in February, another in Pittsburg, Kansas, in the summer—attracted large numbers of farmers and rural families. Once, many of the exhibitors might have shown their products at state fairs, but times had changed. There were other events and distractions at state fairs now, and attendance had slipped from earlier years. In their place, tradeshows were gaining favor with exhibitors aiming to capture a focused audience with increasing mobility but less disposable time. The success of these exhibitors depended on these demographic trends, which was, perhaps, why there were so many out-of-state license plates in the parking lot.

Del and Caitlin usually attended at least three agricultural tradeshows each year. Zach and Amber accompanied them to almost every one. These events were business, education, and entertainment for Del. He viewed them as an opportunity to learn about new products and to talk directly with company representatives.

Caitlin joked that the Farmfest was like a day-long continuing education course for farmers. "There ought to be a tax break for this," she said.

Later she and Amber left to look for Nathaniel's toy tractor and to visit some of the domestic and household exhibitors inside the pavilion. Del and Zach continued to move up and down rows of colorful and sometimes improbable-looking machinery, some of it so specialized and obscure in function that few outside of agriculture would have recognized a use for it. Tractors had undergone profound transformations. Gone were the underpowered tractors of old with their stiff, tricycle steering, cranky transmissions, and uncomfortable metal seats set deep between fenderless rear wheels that kicked up dirt and dust on operators. The new models were, in many cases, four-wheel drive machines with oversized, lugged front tires, air-conditioned cabs, powersteering, variable-speed transmissions, an array of hydraulic pumps, and sleek compartments that sheltered powerful engines.

In the last fifty or sixty years, tractor technology had evolved in parallel with that of automobiles. Today's tractors had no more in common with those of Del's father's generation than did modern automobiles with now classic cars of the thirties and forties. Their manufacturing sites also were now dispersed around the globe.

One fellow, looking at a bright-red Case-International tractor, remarked to Del, "Used to be you could tell whether it was American or not by the color of the paint. Ain't true no more. They make 'em all over now. Hardly any of these are made here."

"Deere's about the only one left still made here in America," Del added.

Even venerable International Harvester, which had risen from humble beginnings at the dawn of the American agricultural revolution in the late 1800s, had been forced to merge with other machinery makers to survive. So had many other farm-machinery manufacturers, all fighting for survival in increasingly competitive world markets. Harvester had ceded much of its manufacturing and assembly to European sites where they competed with automotive giants like Fiat and Volvo, which also had entered farm machinery markets.

Del paused to look at another row of new tractors with gleaming paint and tires coated with wet black. Then he moved on, searching for other must-see exhibitors on their list. Zach examined some new Case-IH tractors, thinking he might get some ideas useful for his competition pulling tractor.

Del's phone rang. "Where are you guys?" Caitlin asked. "We've got food. Amber and I are over near the tractors, but there's so many! We can't find you."

"Well, what kind of tractors are you standing by?" Del asked.

"We're over by some called Ma - hin - dra," she said, pronouncing each syllable slowly, not entirely sure of the name. "I think that's how you say it."

"Mahindra. Oh, you're close. Just look for the red Case-IH tractors. They're right around the corner. That's where we are."

"But these are red too."

"Well, the other red ones, then."

"There's so many red tractors, maybe you better come look for us."

"Okay, hon. Stay put and we'll find you," Del said as Zach listened with amusement to the conversation.

Reunited, the families paused for sandwiches and drinks behind still more red tractors and a display of red cattle trailers. The fender of one shiny cattle trailer provided a convenient shelf for drinks, hamburgers, and fries. Caitlin said she'd signed up for another year's subscription to *Farm Talk* newspaper. For her loyalty, she'd gotten a free cattle whip, a bright yellow one with a rubber tip. "They give us something every year," she said. "This is the first time they've given out cattle whips. And you better behave, Del Montgomery," she said, brandishing it playfully.

He gave it a glance but didn't comment. He already had at least a dozen, maybe more, stashed in barns and trucks.

The midday break was welcome. Nathaniel spilled some of his juice while he stared wide-eyed at the huge, colorful tractors that surrounded him. When a piece of hamburger rolled out of his little fingers onto the ground, he bent to retrieve it, but Amber intervened just in time, admonishing him to leave it on the ground. Del ate quickly and was restless to see more exhibits.

Even before Caitlin and Amber had finished, he fretted. "We got here too late this year. I shoulda known there'd be lots of traffic on weekends."

Del walked between two lines of farm equipment, looking first at one, then the other. He was familiar with almost all of the hundreds of machines and usually knew how they worked even before he examined them. Zach was equally proficient. Making their way through crowds dominated by deeply tanned men in jeans and bill caps or cowboy hats, they looked for cattle-loading chute exhibitors, but got detained by an add-on device that permitted a tractor driver to operate both a hay rake and a hay baler simultaneously. It was an awkward-looking piece of equipment, one that neither of them had seen before, and one of the only machines on which their opinions differed. After talking with the representative, Del thought it might work pretty well, and he peppered the representative for details, even down to the color of the paint.

Zach, on the other hand, was less impressed. "I see a lot of problems with this thing," he argued. "Too many blind corners in fields, or places with bales in the way, or sharp turns where that thing would be hard to use. You'd end up leaving a lot of hay in the field unbaled with that thing."

Del continued his optimistic assessment but didn't like the color, an exotic hue somewhere between nightclub lavender and magenta.

"Would you paint it another color if I ordered one?"

Coming from a farmer, his question wasn't as unusual as it might have seemed. Farmers tended to be traditionalists with strong loyalties to their machines and their colors. Certain colors were always associated with specific manufacturers. So close were these identities that farmers often joked with each other by referring to the color of the paint of their machinery rather than the manufacturer's name. But agriculture machinery colors didn't extend into the exotic blue-violet end of the spectrum, as this rep was finding out. Del made that clear.

"No, we can't change the color," the rep said, looking nonplussed. "Company won't let us."

"What about if I took delivery of one unpainted?"

"Well, yeah, they probably could do that."

"They should've just painted it plain black or something."

"They wanted a distinctive color. Easy to recognize."

"Guess they succeeded. But it's an awful color."

Zach watched the exchange with barely concealed laughter. For the nearly forty-five hundred dollar asking price, he didn't think it was a worthwhile investment, no matter what the color. He finally decided his dad wasn't entirely serious, just leading the guy on, but to make sure he continued his negative assessment.

"Look how long it is. They'll have to boost the amperage just to run the wires from the baler up to the monitor in the tractor. I see problems written all over that thing."

After they left, Del said, chuckling, "Sure had that guy sweating, didn't I?"

Walking downhill to the livestock pavilions, Del said he needed to purchase another Brangus bull, maybe two, this fall to help with breeding duties next summer. He thought this would be a good opportunity to talk to some of the breeders and look at the quality of their bulls. Only a few of the twenty-five odd breeds of cattle on exhibit would have been familiar to farmers and ranchers a generation or two earlier. Most of the cattle here were exotic breeds with names reading like a string of foreign languages: Belgian Blue, Braunvieh, Beefmaster, Charolais, Chianina, Fleckvieh, Gelbvieh, Highland, Irish, Dexter, Limousine, Lowline, Romagnola, Salers, Salorn, and Senepol, among others. A lot of the bloodlines came from Europe, especially England, Belgium, France, and Italy, where cattle-breeding programs had been in place for centuries. Only a few were breeds widely raised in the United States, and the rest were more novelty than core holdings.

"Don't matter what the ancestry is, everybody's breeding for black now," Del told one of the exhibitors, who nodded in agreement.

"Even the Charolais are black. Used to be Charolais were white. Now everybody wants black. If they're black, they can claim it's got Black Angus blood, and it'll sell better to restaurants."

It was mid-afternoon before Del got around to talking to a rep about cattle chutes. He was dismayed that the price of the one he liked had gone up since he'd talked with a dealer in July at the farm show in Pittsburg, Kansas. He bargained with the fellow, came close to the price he wanted, then left, but returned later and resumed haggling. The rep finally relented and gave him the summer price—close to four thousand dollars.

"If this one's as good as it's supposed to be, it's gonna be the last loading chute I'll ever need," Del said with an air of finality as the family regrouped to leave. "The only other one I've ever owned is twenty-five years old. Got my money's worth with that one."

Zach was pleased with the purchase. "At least with this one we can adjust it for large and small animals, and it's long enough to handle big bulls."

"What do you think?" Del asked, looking to Caitlin for approval.

She laughed. "They all look the same to me. I like the color of this one. It's red. That's the most popular color here today."

While Del had been haggling over the loading chute, Caitlin had gone to watch the stock dog demonstration, an event held twice a day. It was hugely popular and held inside a large pavilion with a dirt floor and a rise of wooden benches along one side. Inside the pavilion temporary fences and gates and alleys had been set up. A trainer and an assistant would then put two hardworking border collies through a series of working scenarios to showcase the dogs' abilities. The dogs were smallish and black and white and had an intense stare and a tendency to do everything from a semi-crouch. Both dogs were extremely responsive to the slightest voice or hand commands given by the trainer, and they expertly controlled a group of yearling calves, herded a flock of panicky sheep that behaved just like sheep usually behave, and finally, a flock of skittish-acting gray and brown geese that huddled unnecessarily close together and bobbed like a quivering collection of hyperactive bowling pins when they moved. To control the wary geese, which were herded through two narrow bridges, into an alley, and finally into a penned enclosure, the two dogs made only the slightest of movements, raising a head, crouching, or staring or flicking their tails to control the geese without causing them to become unduly frightened. Caitlin was impressed with the performance.

Rejoining Del and Zach and Amber by the cattle chute, she announced, "I just saw these herding dogs, and they're amazing. And guess what? Now I know what's wrong with our dog, Crissy."

"What's that? Besides not being too smart?" Del countered.

"No. They said that stock dogs should instinctively keep their head and tail low, and crouch when they're herding. Dogs that don't do that automatically by the time they're a year old, or even younger, will never be good stock dogs. It's genetic. That's what these dogs did. They didn't behave like Crissy at all."

"I hope not. Otherwise, that show would have been a big flop."

"A handler explained that the best stock dogs aren't good people dogs, either. They don't make good pets. They want to be herding or working and don't like lying around with people. They need a focus."

"Well, Crissy's not that great with people. Always jumping up on them," Del said. "She's pretty focused on that. We can't even get her to quit doing that."

By the time the Montgomerys left the festival grounds and located their truck, the trees were spreading long, stippled shadows across the hilly parking lot. Standing amidst the rapidly emptying lot, Del felt pleased with the day. Surrounded by the heady atmosphere of so many new and stimulating products, he said he always learned something. He needed the break. So did the family. Tomorrow they'd be thinking about harvest and weather and grain and cattle prices again, and all the work that needed doing to keep a farming going. No amount of new technology would ever completely overcome those kinds of concerns.

33

October 10
Machinery Auction

The air was crisp with the fragrance of old damp leaves and cattle manure when Del, in threadbare jacket and leather gloves with holes in the fingers, turned his black pickup and stock trailer up a narrow lane, past an abandoned farmhouse, and came to a stop next to an old corral and unpainted barn at the top of a rounded hill. Although the hill was scarcely higher than any other in the area, everyone knew it as Dunstan Hill, named after a former life-long resident. Del rented the farm from a daughter of the family. The farm terrain was rolling, and the soil too thin for row-cropping. Some of the hills were rocky, too, but it was good land for cattle. As he stepped out of the truck into the bright autumn morning with goldenrods and asters and giant ragweeds crowding derelict fences, and unwelcome patches of prickly jimson weeds in the corral, Zach and Harlan pulled in close behind in a second truck and trailer.

The previous night the cow and calf herd had come up to drink at a concrete stock tank by the corral, and most of them were still loitering around the corral when Del arrived. With the herd so close, it was easy to coax them into the corral. It wasn't always that way. When handouts of molasses-laced supplement weren't enough to get them in the corral, Harlan and Zach had to resort to four wheelers to corral them. That was why Del decided, more than a year ago, to get a stock dog to help with roundups. Unfortunately, a cow turned on her, and the dog ran. Since then she hadn't been of much use. Caitlin thought the dog had been too young at the time, but she admitted that the dog had been a disappointment.

Del was just pleased to get the herd, some eighty animals, corralled so easily. It was barely past mid-morning, and they had the cattle ready for the vet who would be coming the next day.

At the homeplace, Del and Zach and Harlan sat down on a mismatch of old wooden chairs and a bench on the back porch and pulled off boots and jackets. Caitlin was in the kitchen getting food ready for dinner. "You boys are early," she called out, knowing full well that Del and Zach were early because they wanted to go to the tractor and machinery sale in Grand Forks.

They traipsed, sock-footed, into the kitchen, and Del could see the stress in Caitlin's face. She'd had a hard time baby-sitting Nathaniel.

"He's been fussy. I couldn't get him down for a nap until a few minutes ago, so I'm behind with dinner."

Harlan wasn't bothered by the delay, but Caitlin knew that Del and Zach wanted to get to the sale and look things over before the bidding started. Del looked at his watch and decided they ought to make a sandwich and skip Caitlin's dinner. Besides, they could eat at the sale.

Caitlin wasn't pleased. "It'll be ready soon enough," she said, "and if you boys don't stay and eat it, you'll just get it warmed up again when you get back tonight. That is if you even get back in time for supper," she added, her voice hinting sarcasm.

She knew the machinery and tractor sales often ran on until eight or nine in the evening, and Del and Zach usually stayed until it was over. Sometimes more than a hundred tractors were auctioned in an evening.

Under a thin blue sky scrubbed of clouds, Del and Zach strode past the wide gate in the chain-link fence that surrounded the auction company's machinery and salvage lots. Acres of farm machinery, enough to fill several city blocks, were arranged in long lines. Another section was devoted to used construction machinery.

"Turnout looks light today," Del observed. "Still a lot of guys with beans in their fields, so they won't show up today. A couple months and this place will be packed, and it'll be harder to get a bargain."

Zach nodded, squinting into the sun as he walked past a row of battered brush hogs that looked like they'd mowed most of America's highway roadsides several times. Crockett's Machinery had held a sale here the second Monday of every month for more than twenty years, and their business had prospered, mostly through word of mouth. Now it was one of the biggest outlets for used farm machinery in the Midwest. Buyers came from as far away as Canada and Mexico. Company buyers fanned out over an equally large area, scouring the countryside for machinery from farm sales, dealers who'd taken used equipment on trade, and from farmers and construction companies that brought equipment to sell on consignment. Once on the sale lot, the equipment was sold outright, or at the monthly auction, or towed to their salvage yard for parts.

Zach headed for the salvage yard beyond the sale lot to look for a drag chain and sprockets for an older model John Deere combine that belonged to his grandfather, who was having a hard time repairing his machinery. He walked up and down several rows of rusting combines that were mostly picked clean of usable parts. He found a half dozen like the one his grandfather owned. Only one had usable parts. Returning to the auction grounds, Zach spent the next ten minutes trying to locate Del, finally calling him on his cellphone for directions. "One's got sprockets in decent shape," he said. "The others are worse than Granddad's."

"Check with the guys inside," Del advised. "If the price is right, go ahead and take them."

Inside the large, flat-roofed office and parts building, which separated the sale lot from the salvage lot, Zach leaned against a parts counter that had seen its share of greasy overalls and rough, calloused hands, and discussed prices with the parts manager. They agreed on a price that was less than half of new parts. Zach didn't have his tools but got the manager to send a mechanic out to remove the sprockets and chains. He'd have them by the end of the day.

Even from inside, Zach could hear the auctioneer's stentorian banter over the loudspeaker. Outside, his loud, rapid-fire chant echoed across the lot with a kind of hurly-burly seduction that drew men to the gravitas of the moment. That, of course, was the idea, and rough-dressed men in emblem-embossed caps flowed around machinery like amoebas in flotsam as the auctioneer chanted and cajoled and kept a brisk pace that gave bidders little opportunity to reflect. The auctioneer stood inside a tall camper shell on a pickup truck. With one side of the camper open and shaded by an awning, he looked like an animated puppet in a window. The auction had begun mid-morning with a miscellany of farm equipment, mostly small one-of-a-kind items. It was underway when Del and Zach arrived. By midday the selling would shift to industrial equipment—bobcats, high-loaders, backhoes, dozers, dozer blades, dump trucks—most of which looked shot, like a defeated army wearing worn-out hoses, bald tires, and battered

Caterpillar-yellow paint. During the afternoon, the auctioneer would turn to larger and more expensive farm machinery. The big show was the tractor sale, which usually started about five in the afternoon. Over seventy tractors were up for auction.

The auctioneer was still selling smaller equipment—front-end loaders, blades, scoops, hay rakes, planters, cultivators, even a few moldboard ploughs—and the crowd following him was thin. Men were dispersed around the lot, looking at machinery, making up their minds if something would be worth a bid, thinking about how much they would be willing to give.

Del and Zach watched the bidding, then wandered off to look around the sale lot. Here and there they spotted implements that seemed in good condition, but much of it was high-mileage machinery that had been over a lot of fields. They stopped for a closer look at a tandem disk. "Look at this disk here, all worn out, scrapers sticking out behind the blades. Hydraulics shot too. Take three or four thousand dollars to fix it up, and you'd still have an old disk," Del complained.

An International Harvester combine header caught Del's interest. "Pumps and hoses look good. If it don't bring too much, I'll bid on it for spare parts."

They walked around it a couple more times, looked for dings and rust, flexed the hydraulic hoses, checked for seal leaks on hydraulic cylinders. Satisfied that it had enough usable parts, Del said, "We've got a couple of these already, but we've cannibalized them so much we could use this one too."

It would be a while before the auctioneer got to the header, so they walked through the rows of tractors—a gluttony of tractors new and old, large and small, a potpourri of mostly sun-bleached greens, blues, reds, oranges, yellows, and some oxidized to little more than rust color. There were a few immense, eight-wheel-drive, swivel-hipped behemoths known as articulating tractors. Until a decade ago they weren't often seen in this part of the country, but they were now showing up frequently on larger row-crop farm operations. More common were big, muscular tractors with dual rear wheels that towered as high as a grown man's head. A few of them also had front wheel drivers, tires smaller than in the rear but still heavy and lugged, and these tractors had a burly, brutish appearance almost as formidable as the larger articulating tractors. There were many small tractors, too, mostly older ones now more suited for mowing lawns and roadsides or tilling urban gardens than working on modern farms. Little more than a generation ago, however, these tractors had been the workhorses of American agriculture. Many of the men at the sale today had spent their youth on these tractors, and they still had affection for them. It probably also contributed to the reason why so many of these men bought these tractors and spent time restoring them, even though they weren't essential to their farming operation.

"See that one over there? That big Case-International that don't have a cab?" Del said. "You know that one came from down south somewhere, maybe Arkansas or the Mississippi floodplain. They never have cabs 'cause anybody down there that can afford to buy one of these would never actually drive it himself. They're not gonna spend all that extra money just so some hired driver can be comfortable."

A few minutes later, Zach spotted an old International Wheatland tractor, a W6-TA. Del and Zach already owned one almost like it and had driven it in the Independence Day parade. On the sale lot amidst the newer, monster-sized tractors, it looked like a toy.

"Hey, this thing looks pretty much stock," Zach exclaimed with delight, ignoring the fact that there wasn't a speck of paint on it and it had probably been sitting outside for more than fifty years.

"Still got the hydraulic pump and lever here under the steering wheel," Del noted. "Of course, nothin' fits those couplers now. The ones they made then weren't universal like now."

"Wonder if the torque amplifier works?" Zach asked rhetorically as he pulled back on the lever. "Getting a transmission to drop down a gear without actually shifting was pretty advanced technology in the early fifties." Swinging a leg over the small metal seat, which was positioned low behind the two rear wheels, Zach grabbed the steering wheel and tried the starter. Nothing.

"Check up front, Dad. See if the starter's makin' contact."

"Ain't no spark. It's dead. We already got one like this anyway," Del noted again. "Don't need another, but if it goes cheap I guess we could fix it up for resale."

By midday the crowd picked up; it was not as sparse as earlier. The auctioneer was hitting his stride. Sharp-eyed spotters following the auctioneer's pickup truck, pointed fingers, hollered bids, and jabbed arms into the air in dramatic gestures to encourage bidding. No matter the condition or age of the equipment, it was all introduced with appropriate praise.

"She's ready to go to work for you," the auctioneer would say, referring to each implement in the feminine voice. On the next item he might say, "Help you with those back-breaking chores." If it was virtually worn out, it might be prefaced with a caveat: "Needs a little TLC and she'll be good for years of work." And then the crowd would hear his familiar-style chant:

> "I got sixteen hundred, now seventeen, now seventeen, I got sixteen and a half now, will you give seventeen? I got sixteen hundred and a half, will you give seventeen? Number eleven's got 'er, sixteen-fifty. Pull up, driver. All right now, what've we got here? Looks like a fine John Deere rotary mower."

Occasionally the auctioneer's glib optimism caused heads to shake in good-natured disgust, or provoked muttered wisecracks from the creased, tanned faces of these perennially cynical men appraising the equipment arrayed in front of them. With hands stuffed in overall pockets, one guy leaned forward and spit before passing judgment. "*Shee*-it," he said. "That thing's already plumb wore out."

The auction company grouped similar equipment together so prospective buyers could make comparisons. Disks with disks, field cultivators with field cultivators, and so on. There were grain drills, gravity grain wagons, bush hogs, hay balers, hay rakes, bale spikes, combine corn heads, combine grain heads, loaders, mowers, and augers. The diversity of equipment seemed endless. The auction truck snaked along each row of equipment, pulling forward a few feet, stopping in front of each item. Sometimes two, even three pieces of equipment were positioned one behind the other, and there would be scarcely a pause as the auctioneer moved fluidly from one to the next. If bidding slowed, the auctioneer paused, lowered his voice, and exhorted the men not to overlook a bargain, and maybe added something humorous. Then his staccato cadence would resume, blaring over the loudspeaker mounted atop the pickup camper. Truckers could hear him in the parking lot two blocks away.

If the auction company owned the machinery, there was usually a preset minimum price, especially if it was in good condition. If it didn't reach a minimum, the auctioneer might cajole the bidders briefly, but he soon moved on. They didn't use questionable or dishonest ploys, shills, or phantom bids to protect prices. This was a homegrown business in a small town, and the owner and his sons would not have been so successful if they weren't honest. Older machinery brought in by individuals was typically sold with the designation "absolute" written in chalk on the item. The price was whatever it would bring.

When the auction truck advanced toward the combine header, Del took a position behind one end of it where he had a clear view of the auctioneer. "They all know me, anyway," he said to Zach. "So it don't really matter where I stand. I've been coming here so long I got a permanent number. A lot of these guys do. Some of them buy so much stuff that the auctioneers and spotters even know the auction numbers of these guys. Don't even have to hold up their card numbers."

Bidding was light on the header, mainly another farmer and Del. They sparred with bids, ran the price up to two hundred seventy-five dollars. The other fellow dropped out. Del, pokerfaced, took the nod.

"Good!" he said, nodding to Zach. "We can use this thing, and it sold cheap enough."

Just outside the auction company building, the Knights of Columbus were doing a brisk business at a food concession. Late in the afternoon, Del and Zach bought cheeseburgers and Diet Cokes there. "Tractor auction will be starting directly. I'm going inside, up to the top of the arena where there's a wall. Makes a good backrest. Those are the first seats to go because most of these guys here got bad backs and they'll be lookin' for something to lean on."

Zach went back to look at the old W6-TA again. It was sitting in the first row of tractors, facing the chain-link perimeter fence and alongside other small or older tractors. Several men were gathered around it. Somebody got a mechanic from the auction company to come out with a golf cart full of batteries to try and start it, but he said it was an old six-volt system. He was afraid his newer, high-amperage batteries might damage the electrical system. They'd tow it into the arena for the sale.

The tractor auction took place inside a back corner of the auction company's parts and sales building. A rear corner of the building opened up wide on two sides with immense sliding doors so tractors could be driven diagonally in from one side, stopped in front of the auctioneer, and then driven out the other side. The auctioneer plied his trade from an elevated podium behind the tractors and just in front of the corner support of the building. This was second-hand tractor haute couture, a sort of fashion show runway for used tractors. Each one was towed in, or driven in rumbling and smoking, and paused for perhaps half a minute on the cracked concrete floor. Here they were in full view of an audience of circumspect men with poker faces and parsimonious habits who filled a semicircular set of risers. Almost all of them knew a lot about tractors. While bidding was taking place, everyone could see the tractor and hear its engine running, but each one was soon driven out, usually with bidding still underway, and another prepped to enter. A few tractors with old diesel engines and badly worn injectors spewed clouds of sooty smoke that hung in the overhead trusses and threatened to obscure onlookers in the upper rows, but no one seemed to notice.

Tractor sales were "ride and drive" if the final bid was over three thousand dollars, so a buyer got the right to drive it later and, if he didn't like it, he could refuse the sale. A lot of guys had already started the tractors outside, and some had even driven them up and down between the long rows so they already had an idea what they were buying. It got pretty noisy out there at times.

"Can't tell much about the condition of a tractor motor by listening to it," Del said to Zach. "Even when they're nearly wore out they can still sound all right. Better look for leaky seals, worn hitches, wear in the driver's area, and exhaust smoke."

Tractors were paraded through the auction at a brisk pace. Many of them sold for between three and six thousand dollars. A pretty little John Deere 520, an older model with fresh green and yellow paint, sold for thirty-one hundred dollars, a blood red International 666 brought fifty-three hundred. Tractors were selling well. When the rusty International W6-TA was towed in, the auctioneer listed it as a 1954 model. Zach thought that sounded about right, maybe the only year it had a torque amplifier.

"I'd give a thousand for it, but I bet, with so many guys looking, it will go for a lot more."

It ran up to three thousand dollars in no time, then bidding stalled. Somebody bought it for thirty-one hundred fifty.

Zach shook his head. "Jeez, that's too much. I couldn't resell it for that."

Shadows lengthened behind the auction arena and soon stretched to the far side of the machinery lot. They hadn't gotten to the big tractors yet, and the crowd was larger now. It looked to Del like a lot of guys, some with their wives, had arrived just for the tractor sale. The arena was packed, a sea of jeans and bib overalls. Dozens of farmers were standing at the side of the risers or outside where they could hear the selling. Two men leaned on crutches—one had lost a leg. Another nearby was missing part of an arm.

Several Mennonites milled around outside, distinctive in straw hats, bib overalls, and Abe Lincoln beards. Earlier they had been looking at the tractors, especially the big ones. "Curious that they won't drive anything with rubber tires, even a tractor," Del said, "but they'll buy them and get somebody else to drive them. Some of the men own truck repair shops, too, but they don't drive the trucks."

"I guess it keeps everybody from running around and getting into trouble," Zach remarked.

Del wasn't so sure. "I've heard some of the young'uns are pretty wild. Lots of ways to get in trouble without a car."

It was nearly nightfall. The parade of tractors continued, each now sold under a flood of bright lights that dramatized the blue miasma of diesel smoke and fumes that twisted upward as engines idled under the auction roof. Outside Del noticed another parade developing. Flatbed trucks and semis were pulling up in a line alongside loading docks. The batter of truck motors and rows of colorful running lights on cabs and trailers in the gathering darkness lent urgency to the work at hand. Watching two high-loaders hoisting machinery onto the trucks, Del nodded to Zach. "If you got your granddad's combine parts, maybe we better go. We've already missed one of Caitlin's meals today."

34

October 14
Video Auction

Western from top to bottom in white hat, scuffed boots, and blue jeans, Clint Holland looked like he had just stepped out of a Marlboro commercial. He was large framed and sandy haired, and his weathered, ruddy complexion betrayed years of work out-of-doors. When he spoke, he formed his words carefully. "Just drive real slow across the field and back again in your truck," he explained. "Maybe toss out a few pellets of supplement to get them to follow and spread out."

"Where you gonna be for the filming?" Del asked.

"I'll be over on the side so I can see 'em good and get some video. I been in a lot of cow pastures over the years," he said, laughing easily, "so this isn't too difficult. All I need is about five minutes' worth. Mainly I want to show the calves real good."

Clint worked for Superior Livestock Auction, a Texas auction company specializing in selling cattle over nationwide satellite television and on the Internet. Videos of cattle for sale were posted for previewing on the Internet and shown during live auctions, with bidding carried out via phone or Internet. At the time of the auctions, the cattle were still on farms and ranches, so buyers rarely saw the cattle live. Nevertheless, the nationwide scope of the auctions was an advantage to both buyers and sellers.

Del drove slowly across the pasture, then back again with cows and calves stringing along behind the truck and hoping to be fed. Satisfied with his video, Clint waved his arm to signal that he was finished. He tried to pick out, on video, as many older steers and heifers as possible, those weighing close to 500 pounds. Once delivery was made, which might be up to two months after the video was made, the steers and heifers would be sent to commercial feedlots or smaller feeding operations. Calves would be auctioned in lots of about 90 if their weight approached 600 pounds, or around 100 if their weights were nearer 500 pounds at delivery. Either way, the lots would total about 50,000 pounds, near the size of one futures contract. It was a guessing game looking at calves in a pasture in September or October and estimating what they would weigh at delivery a month or two months later. But it was one in which Clint was highly skilled.

After Clint finished his video, he pulled his pickup truck alongside Del's, the two facing in opposite directions, and placed a seller contract on a clipboard. Most of the information was already filled out, neatly typed in black, but there were details that needed to be settled. How many calves would be delivered? What kinds of vaccinations would they have? And, of course, their estimated weight.

"They're gonna look good, Del. We got a nice tape of them."

"Most of these calves were born in February or March," Del said. "By late November, in a normal year, heifers are about five hundred and sixty pounds, steers six hundred, but it was awful hot and dry this August. Weights might be down a bit."

Clint studied a copy of last year's contract on this same herd. "Looks like your heifers weighed about that last year, but you kept them until December. This year you're wanting to let 'em go a litter earlier. When you weaning them?"

"In about three weeks or a little more."

"Okay, I'll put down weaned three weeks, so I better subtract about forty pounds from the weight since you're selling them earlier than last year. I'll put the heifers down at five hundred and twenty pounds. That okay?"

"Sounds about right."

"But remember, if they're more than ten percent heavier than what you put down here, they'll dock you a little. What kind of vaccinations you giving them?"

"We're planning on a seven-way blackleg, five-way Lepto, and five-way Pasteurella."

"You don't have to give 'em the last one if you don't want to, but it might help if you did. Costs about $3.50 per head extra."

"We're planning to. Helps when you sell them."

"And they'll be wormed and poured?"

"We always do that. We worm in the spring too."

"Where you going to weigh them?"

"Co-op's got a long scales. Rock quarry's further and it's only a short scales."

"I guess to go by the book, we need to weigh them on a long scales so you get all the axles on the scales at once."

Del nodded in agreement.

"What should I put down for horns? Some with horns?"

"Same as last year. They're almost all from those polled Brangus bulls. A few heifers got calves from a longhorn bull."

"Okay, I'll just put down a few with horns. Guess that about wraps it up. If you want to watch 'em sell, it will be on satellite television in a few weeks. I'll call you with the exact date and time, but you can probably get all that on the Internet."

The dry autumn air was redolent with the pungent smell of ragweed and old grass and dust. Del noticed several turkey vultures teetering overhead, their wings fixed in a dihedral, their internal compasses pointing south. Today they were soaring over midwestern farmland, but they weren't stopping. Most of them would make their way southward through Central America, squeeze through Panama's hourglass waist, and be over cattle ranches in northern South America by mid-November, a month-long journey during which they would not eat. By mid-February, having regained their weight, they would begin the long journey northward again.

Del met Zach and Harlan and Caitlin for dinner at the homeplace, the four of them sharing a rare, relaxed midday meal without deadlines. Harvest was finished for the moment, only the late soybeans remained, and those wouldn't be ready for another week or so. There would be plenty of work in the coming weeks but, for a few days, the Montgomerys could look forward to attending to odd jobs around the farm.

After lunch, a machinery salesman stopped by the house. Del sent him off to talk with Zach, who was already outside. "That guy's got a tandem disk he wants to sell," he said to Caitlin. "It's pretty low on my list of priorities right now, but Zach thinks he'd like to try it. It's too big for our regular tractors, but probably too small for an articulating tractor."

Del came outside after the salesman left. "What'd you think?"

"I think it's too big, even the Magnum. But if we could get hold of that articulating tractor that was in the road accident over west of Ashton, then we could use it. You heard anything more about that deal?"

"Still tied up in litigation, last I heard. Lawyers got to sort things out, but it's been a few months since I talked to the guy that owns it. It's got a broken rear axle and some other damage. I seen it over at his place when I talked to him. It would take some work to get it running. If he gets a settlement, I bet he'll let it go cheap."

Zach's face brightened. "I'd love to get hold of that tractor. Got a four hundred horse Cummins engine in it, right?"

Del was thinking the same thing. Maybe he'd give the guy a call this evening.

35

October 24
Last Harvest: Soybeans

Cornflower-blue asters and yellow goldenrods still dotted roadsides and green fields, but the sun was edging southward now. With each passing day, there was a little less autumn to spread across the countryside. The air had the scent of old leaves and soybean stubble and dust. Flocks of red-winged blackbirds wheeled and twisted and settled nervously in freshly harvested fields as Caitlin walked with Crissy along the gravel road north of the homeplace in the evening. Harlan stopped by the house after work to get paid. He said he hated to see winter coming, said he could feel it in his bones, but he didn't elaborate. Caitlin thought he was hungry and gave him part of a loaf of fresh-baked sourdough bread to take home.

The grain driers had been on for nearly a month. With fans screaming, the early milo seed had dried down from twenty-four percent to fourteen percent moisture, a safe level, but the electric bill had soared. Del figured it would be over three hundred and fifty dollars just for the first two weeks of October. He also hoped to finish harvesting the last of his late-planted soybeans by the end of the week. Otherwise, he'd have to interrupt the harvest to wean calves because the first batch would be sold via video auction in a few days, and then trucks would be arriving to load out the calves. More trucks would be arriving by the middle of November for the other groups of calves.

There had been some rain the week before, but after a few dry days, Zach and Harlan were back in the fields to restart the last soybean harvest. At first the heavy combines cut parallel ruts running back and forth through the fields. Del hated to cut up a field. But most of his neighbors' soybean fields looked the same. Some guys were falling behind schedule. Driving back from the co-op the week before, Del saw water standing in deep ruts where one of his neighbors had tried to get into a field prematurely. Another he heard about had buried a combine up to its belly in mud and had to drag it out backwards with a Caterpillar dozer. Now even low-lying fields had dried some, and guys were back in their fields.

There had been a light breeze the previous night, and the morning dawned clear and warm with no dew. Zach and Harlan left early to service and refuel the combines parked on rental land several miles to the west. They started harvesting by eight-thirty. Del had two truckloads of soybeans from the previous day. He took one to the co-op in Prairie Point, but by the time he got back to the field, Zach and Harlan had already filled the empty truck in the field. Del switched trucks and headed back to the co-op again. On the way he called Caitlin and asked her to drive the blue tractor-trailer parked in the driveway to the co-op. "I'll meet you there," he said.

"Will it start okay?" Caitlin asked.

"Should be fine. Just keep the RPMs up. Remember, cab tends to bounce some when it's got a full load and you first start out."

Earlier that morning, Del had called some grain buyers in Kansas City. They were paying $9.27 a bushel for soybeans, the local co-op $9.15, a difference largely offset by trucking costs and time. With harvest running late, he was more concerned with getting his crop out of the field than making a few extra pennies on a bushel. Up north, if wet weather kept farmers out of fields in the fall, they could wait a few weeks until the ground froze, then resume harvesting. That was something he couldn't count on here. Sometimes the ground didn't freeze solid until late December or even January. By that time he could lose some of the crop.

Before Del and Caitlin left the co-op, Zach phoned. "Truck is full again."

"Are you close to finishing the field?" Del asked.

"Not much left. Less than a truckload, I guess."

"Tell Harlan to start moving his combine over to the Hyatt place. That ought to keep him busy for a while, and he won't need a truck. Give us time to get a truck to you."

Back at the field with Zach, Del switched trucks for the third time, and he left the tarp on the truck's trailer rolled back until he got to the highway so some of the chaff would blow off. It might make the grain look a little cleaner for testing.

Del was at the co-op when Zach phoned again, this time with a mechanical problem. "Header reel's not turning," Zach said. "I'm just ten minutes from finishing, but the stems won't feed into the auger without the reel turning."

"You check if the belts are tight?"

"I tightened the belts on the side."

"Keep looking around. I'll call Merle over at AC's in Ashton. See what he says. Could be a solenoid or something inside."

Before Del could call back, Zach discovered the problem was a sheer pin. He didn't have a replacement, but about that time Caitlin arrived with their black pickup and trailer to ferry the combine header to the Hyatt farm. Zach borrowed her truck and checked with a nearby farmer who was able to find a bolt to replace the sheer pin. Within thirty minutes, he was back at work.

By the time Del returned from the elevator for the fourth time, Zach and Caitlin had moved the combine and header to the Hyatt farm, made a few adjustments, and refueled. It was mid-afternoon. Caitlin had sandwiches for Zach and Harlan, which they ate in a rush just as Del showed up with an empty truck. Smiling broadly as he bounded out of the truck, he said, "Hey, hon, you got a sandwich for me too?" He hadn't eaten since about six in the morning. "I'm starved."

She handed him a sloppy joe and a banana. Holding his sandwich and a drink, he checked the reel on Zach's combine and a belt they'd replaced the day before.

While the men finished their sandwiches, Caitlin laughed and said, "Guess what? I caught a mouse this morning. I heard the trap snap after breakfast. It was right after you left the house. Flipped it up into the sink and stunned it, I guess. It was just laying there shaking a little."

"I'm surprised it didn't get caught last night. Mice getting pretty bold walking around in broad daylight."

"Yes, and with a cat in the house too."

Del laughed. "That fat ol' cat probably don't scare anything. She's too slow and out of shape to catch one. Besides, she doesn't have claws anymore."

Listening to their conversation, Zach was laughing as he started up the metal stairway to the combine cab.

Del winked at Caitlin. As he turned toward the big white semi, he said, "Hon, better set some more traps tonight."

All that remained of the more than nine hundred acres of fall harvesting was about fifty acres of soybeans. They'd finish by tomorrow noon. Zach said that as soon as they finished, he was going to help a neighbor with his harvest. Del figured it would be after dark before they quit, same as nearly every day for the past several days. But harvest was nearly over, and there was almost a full moon tonight, a harvest moon, perhaps.

36

November 3
Cattle Auction

November. Spring calves weaned and sold; veterinary continues pregnancy checks of cows; open cows sold and pregnant replacement cows purchased at auctions; best heifers held back for future breeding stock; heifers and newly purchased cows branded; soil testing begins; lime and anhydrous ammonia may be applied to fields in anticipation of spring planting; last crickets call early in month; most leaves have fallen; deer hunting season begins amidst much anticipation.

The auctioneer's chant was fast, his staccato baritone resonating across the enclosed arena like a hypnotic oratorio. "Got a short and solid here, third period, bred to black. Hip, I got nine hundred here, waddya gimme, nine twenty, nine twenty, waddya gimme . . ."

Buyers flashed stealth bids in rapid succession. A spotter pointed and peppered the fast-paced action with barking hips and yelps, jabbing his finger out with each new bid, his eyes roaming over the crowd. In the cattle pen below the auctioneer, two drovers slapped their whips hard again the white-painted steel piping of the corral to keep the pod of cattle moving so buyers could look them over.

In less than a minute, the auctioneer repeated his final price once, then twice, and with scarcely a break in his chant, intoned the words, "Git 'em outta here," a cue to drovers to move the cattle out, and to prospective buyers that the selling price was about to close. He took a nod from the last bidder and the sale was done. It happened so fast that there was little time to think.

A heavy scuffed door on the right slid open, the drovers' whips cracked, and the small group of cattle were ushered out. In seconds the uncomprehending group vanished into the dimly lit wilderness of cattle pens outside.

By the time the cattle disappeared, they were already an entry in a computerized spread sheet, one more knot of cattle moving efficiently through the labyrinth of holding pens and alleys that led to waiting trucks at the far end of the auction yard. Most sales were over in less than a minute.

Del and Caitlin, along with many others, arrived late, all of them standing in line to register and pick up buyers' cards before entering the sale arena. A polite sign at the cashier booth entreated customers to please be patient because the tellers were learning a new computerized system.

One man, shoving his hands in his pockets and shrugging at the long wait, said, "Hell, a guy could git hisself throwed in jail quicker 'n you can git a number here."

When Del finally got a card, he was buyer number 135 to register for the night. The number indicated good buying interest, but there would be larger crowds in coming months. "Some of these guys are still harvesting corn and soybeans. Haven't sold their crops yet, so they won't be buying for a month or two," he said to Caitlin. "Most of 'em haven't got their calves sold, either, so money's still tight."

"Oh, with you farmers, money's always tight," she said with a mischievous smile. "Maybe just a little tighter than usual tonight."

Del ignored her teasing. "A lot of these guys won't sell their calves until mid-November or later. Then replace nurse cows after they get rid of the nonbreeders and figure out how many replacements they'll need."

Caitlin picked up a buyer's card too. Del admonished her with a grin. "If you buy something here tonight you're on your own. I can probably spend enough money for the both of us."

"Oh, you never know. I just might see some I like," she said, continuing to tease him.

Del was already moving toward the arena, climbing the stairs to the seating area. Caitlin followed, and they found two empty seats together high up in the center of the arena. They leaned back in the well-worn, red-vinyl seats with cushioned, fold-up bottoms that would have looked at home in an old movie theater when most of the men and women there were teenagers. The auction was already underway, the auctioneer's steady thum soaring above an undercurrent of hushed voices. The crowd looked relaxed, but the nonchalance that seemed to afflict the audience was deceptive, more studied stoicism than inattention. Most people, in fact, were paying scrupulous attention. They had to. Selling was fast, and prices sometimes got bid up quickly.

"I'm hoping demand won't be too high tonight," Del said in a soft voice to Caitlin. "Sometimes cattle prices get higher late in the fall and winter."

Caitlin nodded as she looked across the arena.

"More guys around with money in their pockets then."

Caitlin, like Del, knew that cattle prices were unpredictable. It was hard to predict which way things would go. Del had called the auctioneer earlier in the day to ask about prices for bred cows. He'd told Del he thought bred short and solids and younger cows would sell in the eight- to nine-hundred-dollar range. Top quality a thousand and up, higher than a few months ago. That was more than two hundred dollars higher than a year ago. It seemed expensive, but Del remembered that he'd gotten a dollar fifteen to a dollar twenty-five a pound for his calves—the highest price he'd ever received. He told the auctioneer he might be late getting to the auction. Did he know when most of the bred cows would be sold?

"As long as you're there by six-thirty, you won't miss anything important," the auctioneer told him.

It was nearly seven by the time Del and Caitlin got their buyer numbers and found a seat, but drovers were still bringing in older cows, mostly singles and pairs, some with calves at their side. Once they ran through these, they'd move on to bred cows, grouped according to quality. Heifers and bulls would sell last.

"The guy was right," Del said, leaning toward Caitlin. "We haven't missed anything."

"Nothing much selling now that you'd want to buy and keep," Caitlin said, her eyes wandering across the crowd, looking for a familiar face.

Del relaxed. It was warm and comfortable inside the well-lighted arena. Looking around, he figured there were at least two hundred and fifty people, buyers, sellers, and onlookers, here for business or a little entertainment. The seating arena wrapped around the white corral in a 180-degree arc. Each row of seats rose steeply above the preceding, like risers in a stadium, offering an unobstructed view of the cattle in the corral below. Cattle were ushered in on one side of the corral, displayed briefly, and then herded out the opposite side. If bidding was brisk, it continued for a few moments even after the cattle left the arena. Two auctioneers and two recordists, facing the

audience, sat in a wide booth behind and above the corral. An electronic signboard to the left of the booth showed the number of cattle being sold, their average weight, and total weight of the group in large red numbers. On the right side another digital sign showed the number, weight, and final selling price of the last group of cattle sold. It was everything buyers and sellers needed to know.

The crowd was composed mostly of farmers and cattlemen in denim work clothes and caps. A few had on well-worn cowboy hats. Collectively they hardly looked a picture of prosperity. Yet, by the end of the evening, almost a million dollars' worth of cattle could change hands. Some nights the figure would be higher. A few women and children also watched the selling. A scattering of Mennonite families, men and boys in bib overalls and dark shirts, women in long dark dresses and white scarves, sat along one side of the arena. They were knowledgeable buyers. When they came it was usually to do business.

Commercial cattle buyers were always present, usually looking for feeder calves in a specific weight range, or heifers for breeding stock. Others would purchase older cows, the canners and cutters, for slaughterhouses. These men were specialists and they knew their markets. Profit margins were often razor thin, maybe only a penny or two a pound, and they studied cattle prices like a religion. Their cattle buying was dictated strictly by numbers.

Cattle were brought into the sale arena singly or in groups up to about nine. Once inside the high corral, the door closed behind them with the authority of a penal institution lockdown. Using whips if necessary, drovers inside the corral kept the cattle moving around but generally stayed within a step or two of a heavy steel safety barrier at the back of the corral. The barrier was constructed like a false runway with a narrow opening at either end, large enough for a man to slip through but too narrow for a determined cow. This was no job for the inexperienced or slow of foot.

Each animal that entered carried one or two stick-on tags on their back. One indicated the animal's age and identification number. Pregnant heifers and cows got an additional round, color-coded tag indicating the trimester of pregnancy—blue for first, red for second, green for third. Open cows, those that didn't breed or weren't bred, carried a yellow tag. If the cattle were part of a larger herd, buyers were told they could bid on the first batch in the arena and take the remainder at the same price. Some did. Others bid on each batch separately as they were brought into the corral. For his part, the auctioneer gave a fast, telescoped descriptive of each animal or batch. Buyers had to listen carefully.

After one particularly brief introduction, Del told the fellow next to him that he was glad to have Caitlin along to confirm what the auctioneer said. "Easy to miss something he says. You can't stop these guys and ask questions."

"Yeah, and you better be careful where you scratch too," the fellow said. "You might buy a cow."

"Don't want to be bidding on something if you don't know what it is neither," Del said. "You hear about the guy a year or so back that bid on four cows?"

The fellow shook his head.

"It was his first time at a cattle auction. Thought he got a bargain, but when he went to pay he found out the price was for each cow, not the batch. He didn't have the money."

"Sale barns don't like that. What'd they do?"

"Had to run the cattle through again at the next auction. Didn't bring as much the second time, and he had to make up the difference. Had to pay the barn for feeding them for a week too."

Del and Caitlin enjoyed the auctions. Zach, however, had never had as much interest in the cattle-buying side of the business as he did in taking care of them. This evening he was still busy helping a neighbor with a soybean harvest, so attending the auction wasn't an option.

The auction barn usually held a livestock sale every Thursday and occasionally added special sales, and if Thursday was inconvenient there were plenty of auction dates available from other sellers in the bi-state area. Del and Caitlin didn't get to a sale every month, but they tried to go several times a year. It was a good place to see neighbors and friends, too, a social gathering for buyers and sellers.

The auctioneer's jargon was fast and clipped, everything minimized. Almost like a foreign language to the uninitiated. When he said he had a batch of shorts and solids, buyers knew he was selling cows at least seven years old that had all their incisors, although the teeth were usually worn down some, hence short. A broken mouth was a cow with missing teeth, almost always an older cow—not a good sign if you wanted breeding stock. Bred to black was a cow bred to a black bull, by implication an Angus or Brangus bull, but breeding information was often sketchy. Heifers and bred cows less than seven years old were usually indicated by age if it was known. Young animals destined for feedlots and old animals usually sold by the pound. Pregnant cows and bulls sold by the head.

Del liked the auctioneers at this sale barn. "They make sure you can see the cattle from all sides. And they point out stuff that might affect the price too. Long toes, a bad eye, a limp. Some auctioneers make light of a defect because it will hurt a sale," he told Caitlin.

Caitlin was skeptical. "Oh, they all go so fast it's hard to get the details even if they do tell you. I wish they weren't always in such a hurry."

"If they're selling your cows, that's what you want," Del countered, keeping his eyes on the auctioneer below.

In an era of electronic overkill, the absence of hand-held devices for computing prices and numbers was striking. Most buyers favored a pencil and a few 3 x 5 cards ruled for price, weight, and number of head on which they jotted down purchases. A few noted purchases in small notebooks. Only one buyer, an older man with gray hair, jeans, and a red plaid shirt, held a laptop. Sitting to one side of the arena, he watched the selling intently and occasionally tipped his silver-framed glasses forward to study the small screen. Like most buyers, his purchases were discrete. Del suspected that laptops and other electronic devices would eventually show up here, too, a realm where traditions didn't die easily.

At the back of the arena, on the left and halfway up, a door exited to a long, elevated metal walkway outside. The walkway extended out over the tops of the corrals and provided a dramatic bird's-eye view of the cattle and holding pens below. From this elevated T-shaped walkway, prospective buyers could look down and study penned cattle before the sale. An earthy smell of cattle and manure and pulverized soil permeated the area. The stockyards were a sprawl of corrals and alleys as big as a city block, all of it covered by a metal roof. Cattle pens were built with heavy, unpainted pipes oxidized and stained brown from years of cattle and dust, and from lean, hard men who climbed on these piped corrals and moved cattle through them. The corrals were strong, like an endless series of gridironed frames locked in an iron embrace. Fine dust filled the air, a colloidal-like suspension kicked up by thousands of hooves that pawed and churned the dry pens and alleys. The dust, like a dingy haze, dimmed the amber lights illuminating the corrals and imparted an almost surreal atmosphere to the men and cattle and goings-on below.

Some two-dozen men on horseback and on foot shifted batches of cattle steadily through the labyrinth of corrals and swinging metal gates, sorting and moving them like living, larger-than-life pieces on a giant chessboard. Some corrals were full, others empty or with only a few animals. Cattle waiting their turn in the selling arena were held toward one end of the stockyards. Group by group, they were shifted ever closer until only a heavy steel door stood between them and the bright lights and noise of center stage and the auctioneer's chant. Sometimes bewildered cattle ran in the wrong direction and attempted little rebellions, but the cracking of drovers' whips and strong gates inevitably restored order.

Once sold, cattle were quickly shifted back through a series of runways and corrals to the opposite end of the yard for loading. Others were detoured past an assembly line of efficient veterinaries and record keepers who administered vaccinations, performed tests ordered by buyers or sellers, and entered the history of each animal into a computerized database. It took only minutes.

The cattle pens extended off into shadows at the far edge of the stockyards, where a long line of cattle trucks waited in the night. Their red and orange running lights illuminated the darkness beyond the yards with flecks of color that contrasted with the monochromatic jumble of brown corrals and brown earth. Diesel engines idled, churning small twists of unseen exhaust into the night sky. An occasional yell or whistle or crack of a long whip echoed from the semi-darkness as clots of cattle were hurried along alleyways and into holding areas. Heavy iron gates opened and clanged shut. Approaching the trucks, coalescing lines of uncomprehending cattle flowed like lemmings down alleys and upward into the bellies of cattle trailers. By morning, most of them would be gone, trucked into the night, some crossing two, even three state lines by the next day.

Cattle entered the sale arena in various ways, some calmly, others cautiously or with their heads down and feet splayed as if ready to flee. Some charged boldly into the arena expecting an escape, only to circle in bewilderment with heads held high. Over the years Del had learned to avoid cattle that acted wild and restless in sale arenas. He studied the behavior of each group closely because they often behaved the same way on the farm, jumping fences, stampeding recklessly, charging when they had calves. It could indicate problems he didn't need.

Del watched the selling of several groups of eight or nine cows with calves. Some of the calves seemed confused, tried to stay close to their mothers, but sometimes got underfoot as the animals paced back and forth, excited by the noise and lights of the selling arena.

"These will probably be the highest-priced cattle of the night," he said to Caitlin. "Good quality. All from the same herd."

The auctioneer told the crowd the cows were from someone who was retiring from the cattle business. He mentioned the man's name, a gesture aimed at reassuring buyers, and a few of them surely would know the seller. It could make a difference. The cattle sold high, over sixteen hundred dollars for each cow with calf. Del bided his time. More groups from this same herd sold for nearly as much.

"I'd rather buy a cow with a calf in her, than with one at her side," Del said, leaning toward Caitlin again.

"I know what you're saying. Little calves can get hurt, even trampled, during these sales."

"Dangerous moving cows and calves together in trucks too."

They watched as more bred cows, all shorts and solids, were sold. Prices weren't bid up quite so much, but the shorts and solids were older cows, so the lower price wasn't surprising. When several younger bred cows were brought into the arena, Del decided to bid. He got two at eight hundred sixty dollars each, two more at eight hundred seventy, then four red ones, three and four year olds, for eight-fifty each. He was hoping for more but nothing suitable turned up for several minutes.

He'd figured he'd seen the last of the younger bred cows when two black ones were brought in. They looked nice but somebody else also wanted them. The bid ran up to nine hundred. Then the guy stopped. Del had, in his mind, an upper limit of nine hundred dollars too. He waited, then at the last moment, bid nine hundred and ten dollars. He told Caitlin that if you didn't have an upper limit in mind, you'd often get caught up in the bidding and pay too much. He'd gone just a little over his limit but figured the extra ten dollars apiece was worth it.

Three teenage boys were huddled together on the top row of the arena, telling stories punctuated with animated arm movements and bobbing heads, prompting the auctioneer to stop at one point and poke fun at them. "Was that a bid I just saw from one of you young fellas up there?" he said. "I think it was. I just sold you that last cow. Did you know that?"

The crowd chuckled and the three looked sheepish.

Del would have bought more, but the auctioneer moved on, selling heifers and young bulls. He wouldn't see anything more to bid on tonight. The ten he got would make a small load.

After queuing up at the business office to write a check for the cattle, he walked around to the loading area in the back and asked one of the on-duty vets to work the cows tonight—vaccinations, ear tags, and a pour-on drug for worms and other parasites. It would cost more than if he did it himself, but he didn't have the time. This way, they'd be ready to turn out to pasture.

When Del and Caitlin left the sale barn they felt pleased, having at least gotten some replacement cattle. They weren't always so lucky. Sometimes they'd go to two, even three auctions to get a load. It was dark outside beyond the sale building. The unlighted parking lot was almost empty now except for the line of trucks waiting to load out at the far end of the stockyards. A few pickups, their headlights flashing cones of light into the darkness, seemed to move randomly across the parking lot, then coalesced into a line that swung eastward toward the highway. Caitlin glanced up. The inky sky was stippled with a thousand pinpricks of starry light. Pegasus, the great winged warrior horse, was already well along on his nightly journey through the autumn sky.

Thinking out loud, she said to Del, "I guess there aren't any constellations named for cows?"

"Isn't there a bull up there or something?"

"Well, you might know it'd be a bull, not a cow."

"Don't much matter," Del replied. "It's all just somebody's imagination anyway."

37

November 10
End of Innocence

Zach, Del, and Harlan stopped to rest for a moment. They'd been weaning calves since shortly after daybreak. Working on foot inside a corral, they patiently culled calves from cows by opening and closing a long swinging gate. They moved slowly among the milling herd, maneuvering a few animals at a time into positions where calves could be separated from their mothers. Their whips made sharp cracking sounds and sometimes they yelled and moved quickly to blunt the trajectories of stubborn cows.

Leaning against a rough, unpainted corral post to rest, Harlan said, "I seen a big buck last evening. It was at the back end of that pasture over yonder where we were rounding up these cattle."

"You don't see them out by day much unless there's a doe in heat," Zach added.

"With deer season starting this weekend, everybody's out scouting around. Could be somebody scared it out of the timber back there."

"You going huntin' this weekend?"

"Gonna try. Least ways I am if I ain't still chasing these dang cows." Harlan was excited about deer season and talked rapidly. "I had to run some guy off my property a couple days ago. It was almost dark. All of a sudden I heard a shot. Made a beeline straight over there and told him this was private property. He told me that John, my neighbor, said he could hunt here."

"Maybe John just meant his own land," Zach interrupted, "or the guy didn't know where the property line was."

"Well, yeah, I'm gonna have a talk to John about that. Hell, that guy had on all dark clothes, no orange. Man, that's dangerous. He could get hisself shot."

"Wasn't hunting deer out of season, was he?"

"He said squirrels, but I bet he was checkin' it out for deer."

Zach had deer hunting on his mind too. He had been debating whether to go deer hunting this weekend or enter a tractor pull in Columbia, Missouri. He liked to hunt on the first weekend of deer season, before the deer got spooked. So did everybody else. He had also wanted to go to the pull until he found out they were requiring Kevlar scatter-blankets around the tractor's flywheel and bell housing. He didn't have one because they weren't required at regional meets.

"Easy decision," he said. "I'm skipping the pull and going deer hunting. I'm going to sell that tractor this winter anyway."

Zach had been hedging his bet because he'd already negotiated his deer-hunting strategy with Amber, who'd also entered the deer hunting equation. She surprised him four years ago when she bought a .243 Winchester, then bagged a deer on her first hunt. She hadn't gotten one since, but that hadn't stopped her from trying. Zach agreed to baby-sit the first weekend while she hunted. Then he looked at Harlan and grinned. "Well, I'll probably get in an hour or two in the evenings when she gets back. That might be the best time anyway. She usually doesn't stay out that long."

"I still don't know exactly where I'm going this weekend," Harlan said, hoping Zach might offer to let him hunt some of their rented property. Zach didn't take the bait. He knew Harlan wanted to hunt some of those properties, but he wanted them, too, and he didn't know if the landowners' generosity would extend to hired help. When it came to deer hunting, Harlan would have to fend for himself.

"How about your own place? You got a little timber."

"Yeah, but it ain't very big. I heard people say deer huntin' is gonna be harder this year. They say there's lots of acorns in the woods. Then the deer don't come out in the open much. Makes it harder to get one."

"People always say stuff," Zach countered. "Don't always make it true. Besides, there's not many oaks out here on the prairie, and there's deer here."

It was after nine in the morning when Del and Zach and Harlan finished weaning the calves, but their work wasn't over. Now they had to sort the calves into three batches. The largest group, those born in February and March and weighing a little over five hundred pounds, would be separated for sale in a week or two. Then they separated three large ones—calves too big and heavy for the video auction. They had been born earlier and would have to be sold at a local auction. Finally, there were a few small ones, born in the late spring or summer and too young to wean, so they'd go back to their mothers.

Calf weaning marked the end of a calf's dependency on its mother, that fateful moment when the biological connection between cow and calf was ruptured and the beef industry stepped in to take control. The shock of weaning was probably temporary, a trauma likely no worse than spring roundups, which included castration and vaccination. It was something that would happen naturally when a cow's milk dried up, and was bound to occur fairly soon anyway. After weaning, the lives of these calves would forever change. If their first endless summer nursing and grazing and acting frisky in a pasture was the equivalent of childhood innocence, then the next eight months or more would be a rapid transit into early adulthood, and to a diet and life utterly unlike anything ruminants were evolutionarily designed to face. Feedlots emphasized weight gain through high-caloric diets and minimal exercise. They also would likely get growth hormones until the last month or two before they were sold. The upside, if you were a steer or heifer, was plenty of food and water and the security of round-the-clock medical attention. The downside of a high-protein diet for an animal built to ferment grass in cavernous stomachs was digestive problems, some potentially fatal. There also were health issues associated with crowding animals together in close quarters. Getting large amounts of beef to markets quickly, and at affordable prices, had spawned an industry that was highly efficient, but it was not without controversy.

This morning, however, Del's concerns were about the struggle of wills that ensued as cows and calves resisted separation. A determined or confused cow that weighed twelve hundred pounds could be dangerous, even accidentally, apt to knock down or trample anything in her path. In the melee, cows sometimes rushed past men, who took calculated risks each time they tried to halt a cow's breakaway. Once the nearly two hundred cows and calves were rounded up in a large corral, the men began moving small batches of cows and calves together into a smaller corral for the tedious process of separating cows from calves. After a summer grazing in open pastures, cattle mostly resisted the idea of confinement. Del and Zach held their arms spread wide and carried long whips to control the milling herd. When that failed, they'd yell or dash ahead in an attempt to separate animals. It came down, ultimately, to getting a corral fence and steel gate between nearly a hundred relatively slow-witted cows, who had little comprehension of what was happening, and almost as many equally confused calves. The calves, however, were quicker and nimbler than their mothers, but more easily intimidated. And the corrals were muddy. In the churn of mud and hoof prints in the corrals, men and animals alike slipped and skidded and got splattered with mud and manure. Harlan stationed himself by a red metal gate, ready to open or close it as quickly as he could to let calves run into a smaller corral, or to cut off cows. Sometimes the confused animals rushed past him before he could cut off any calves, and when that happened Del and Zach would circle back and start over again. Sometimes it was hard staying upright and out of harm's way, but all three men were experienced around cattle and understood, almost reflexively, when to give a cow space.

Once, when things weren't going well, Del stopped to catch his breath and said to Zach and Harlan, "You know, a cow's only as smart as she needs to be, and most of the time it's not too hard being a cow." Then, after a pause, he added, "Compared to a cow, you'd think we ought to have an advantage in brains in this sortin' business, but sometimes I'm not too sure."

"You'd think so," Harlan said, "but a cow's a lot bigger than me and she can run faster too. Comes down to brains or size and speed. Take your pick."

Little by little, the calves got winnowed from cows. After that things got noisy. Cows couldn't get to their calves and bawled long, lamenting bellows that started low and rose to a high-pitched trumpeting, as if their voices had gone falsetto. Amidst the maelstrom of noise, the men loaded the weaned calves into two long, black stock trailers. Later they opened the corral gates so the cows could return to the pasture, although most of them stayed near the corrals. The calves had already been sold via the video auction. Now all that remained was for Del and Zach to haul the calves to the homeplace where they'd be vaccinated and held until trucks arrived to take delivery of them.

"Buyers don't like to buy extra pounds on calves," Del told Caitlin during dinner later than morning. "They want to grow the pounds themselves. That's why we got to get the weight right so we don't get penalized for shipping calves that don't meet what's specified in the contract."

That was where Clint Holland came into the picture. When he videotaped calves for auctions, it was his job to help sellers estimate average

NOVEMBER 10. END OF INNOCENCE   337

weights of calves at shipping time. To do this he quizzed sellers about weather and grass conditions and feed supplement, and he looked at bloodlines too. His sellers were rarely penalized.

Clint usually came around on mornings when calves shipped. He'd been out a few days before for the first load of ninety-five steers and heifers that Del sold. He'd estimated that batch of calves would ship at an average weight of five hundred twenty-five pounds each. At the final weigh-in, the steers averaged five twenty-seven, the heifers five nineteen. Clint had been off by about three and a half pounds overall.

With the second batch of calves ready to ship, Clint was back. The calves Del and Zach and Harlan had separated belonged to one of the rental farm owners. It was the only herd that the Montgomerys didn't own outright. They took care of the herd for a third of the sale, with the owner footing the bill for supplement, vet services, and annual replacement cow purchases. Clint came early, as usual, to oversee the loading and monitor the sizes and weights of calves. Buyers wanted uniform batches, and Clint could reject any that were too large or too small or had health problems. One calf was a little larger than the others. Del wondered if Clint would reject it. He didn't. But he studied it for a moment, perhaps deciding that it wouldn't affect the overall average much. After a few minutes, he signaled to load all of them.

"Cattle hauler's at the truckstop out by the highway," Clint said. "Saw him there when I came by this morning. He probably slept there in his truck last night. I'll give him a call and tell him we're about ready to load out. Those guys usually like to get some rest and eat before they load cattle 'cause once they've got a load, they can't stop until they get where they're going."

"These calves going to a feedlot somewhere out in Kansas this time?"

"I think so," Clint said. "They'll weigh eleven or twelve hundred pounds by next summer. Look real nice then. Maybe up to thirteen hundred pounds by slaughter."

From today henceforth, these calves would likely never taste another blade of grass, never see anything but the inside of a fattening pen. They'd share it with around ninety other like-minded, similar-sized animals in a sprawling feedlot where hundreds of identical feed pens were linked together in a kind of bovine Plainsville, a suburbia for weight gain with tailor-made diets. To satisfy the ever-increasing demand for beef, especially tender, near-identical-quality beef, the meat industry streamlined every aspect of cattle feeding to bring animals to market weight in the shortest time possible. Clint was right. They *would* look real nice by the end of next summer. In fact, finished out, they would all look almost identical, weigh the same, and taste the same. This was a high-volume, low-margin cattle-feeding business, and feedlot room and board was expensive, but it got far more cattle to market than would ever be possible on grass—more, in fact, than could even be raised on all of the remaining grass in the country.

Grass had now become a luxury for an animal whose evolutionary roots had been tied to grazing. Feedlots had transformed the eating habits of Americans, turned beef into caviar for the masses, brought a consistency of flavor and quality to the meat that had never before been possible. Beef had become a relatively low-cost household food commodity. That consistency, however, came at a steep price, most of which was not reflected in what consumers paid for a shrink-wrapped T-bone or sirloin at a supermarket. What made it possible was the amazing abundance of corn, and to a lesser extent, other grains such as milo, which also were over-produced and, until recently, chronically depressed in price. Agribusiness loved low commodity prices, encouraging overproduction through lobbies that worked hard to maintain costly farm subsidies that got passed along to taxpayers.

The meat industry, not surprisingly, promoted corn-fed animals as a virtue to consumers, yet the real virtue was to an industry that, through high-carbohydrate corn diets supplemented with other grains, liberal quantities of protein supplement, and low-cost hormones and antibiotics, could grow more beef in ever shorter timeframes. Grass-fed steers needed about twenty-four to thirty-six months to reach a slaughter weight of 1250 pounds; a corn-fed steer could grow to almost the same weight in fourteen to sixteen months, an astonishing difference and one with huge economic implications. But what consumers got with that consistency of flavor and quality was beef with more, not less, saturated fats—more omega six fatty acids, which were not beneficial to them, and fewer of the omega threes, which were. With grass-fed animals, the opposite was true, and that was a powerful argument for change, although one no longer practical unless beef consumption by humans changed.

With health problems on the increase, a lot of fingers had been pointed at beef consumption. What those fingers should be pointing at, perhaps, was not at beef generally, but at grain- and corn-fed beef specifically. Meat from grass-fed animals, which were eating what they were evolutionarily designed to do, was far healthier for humans than corn-fed beef. An argument for more grass-fed animals seems persuasive, but bumps up against economics as well as the fickle tastes of consumers. Slowing the system and increasing the number of animals on grass would almost certainly create beef shortages and cause prices to rise. At the present rate of beef consumption in the United States, there isn't enough grassland available to finish out the number of slaughter-weight animals currently coming to market. This could be offset, perhaps, by increased imports, but that would likely put pressure on foreign beef producers to convert more tropical forests into grassland, and consumers demanding only grass-fed beef would have to be willing to expect more variation in the quality and taste of their meat, consume less beef, and pay a much higher price for it. Would they do that? Ultimately, the choice was up to the consumer.

Twenty minutes later, a blue Mac truck with a sleeper cab and gleaming, double-decker cattle trailer pulled into the Montgomerys' driveway. The driver stopped the truck just beyond the metal barn and backed up toward the cattle chute positioned at the corner of the barn. The new cattle trailers, with smaller circular holes for ventilation, were more enclosed than the old ones, but they weren't designed just to put joy-riding convertible owners' minds at ease. A cow with loose bowels could still ruin a convertible owner's day if he was in the passing lane, but the new trailers were easier to clean and, by limiting what an animal could see, provided a more calming environment.

The men called in low tones to get the calves moving and occasionally yelled encouragement to balky ones, but the aim was to avoid frightening them unnecessarily during the loading. Caitlin watched, standing near the corral. The November calf sales were one of the biggest sources of farm income for the year. This was a payday jackpot. Seeing the calf crop loaded and shipped brought everything to closure, the culmination of a year's work, and Caitlin wanted to savor the moment and the excitement. Del worked the alleyway leading to the loading ramp. He carried a battery-powered hot prod in his left hand.

"Helps them make up their minds," he said, grinning at Caitlin. "Keeps 'em movin' so they don't pile up in the alley. This is crowd control."

Most of the calves readily moved forward, moving into the narrow corral alleyway and up into the truck. Perhaps confused by the noise, a few panicked and ran toward the rear of the corral, banging hard into the metal swinging gate that Zach held. He pushed valiantly to stem the onrush of the little group of five-hundred-pound renegades in full retreat. But in the mud he had little traction and couldn't hold the gate. All he could do was step aside and try to keep from slipping and falling down. Plodding through mud to the back of the corral, he urged the calves forward again. A few ran straight down the alleyway to the loading chute, then balked at the noise of others in the trailer, tried to turn back, and became stuck crosswise in the narrow alley. When that happened, hot prods buzzed and balky animals soon changed their minds. In the end, all of them dutifully climbed the loading ramp and into the din of stamping hooves and jostling bodies inside the trailer. Forty were loaded on an upper deck and partitioned into groups of about ten; forty-four more were partitioned on the lower deck—close to maximum legal weight.

The driver, who remained inside the trailer during the loading, rearranged gates to distribute and restrain the calves. "That all of 'em?" he asked as he climbed down out of the trailer and pulled off heavy, manure-stained coveralls and rubber boots to reveal clean clothes underneath.

Securing the trailer doors, he handed Del a hand-written receipt for delivery, and immediately began backing the long trailer down the driveway, then slowly angled it to the left and onto the highway. By the time Del and Clint reached the back porch of the house, the truck was nearly out of sight, headed to the co-op for a final weigh-in. Within fifteen minutes, he called Clint with the weight results.

Clint sat with Del at the kitchen table to calculate weights and prices and write checks to Del and Zach and to the owner of these calves. Prices were good but, as usual, steers brought a little more than heifers because of their expected faster weight gain. Clint was careful, running through his calculations twice, figuring the average selling weight and price for the steers and heifers separately, deducting an initial down payment, and finally deducting two percent of the total weight for shrinkage during transport before coming up with a grand total. He called his office, gave them the figures over the phone, and asked for a computer check of his calculations. It came to $53,597 dollars before selling commission, down payment, and vet fees. The Montgomerys would receive a third with the balance going to the landowner who owned the cattle. On all other calf shipments, the Montgomerys would receive the full amount.

When Clint finished writing his checks, it was past midday. As he stood to leave, Del glanced anxiously at a clock over the kitchen sink. He had been asked to be a pallbearer at a funeral at two. He figured he might have to skip the noon meal to make it on time.

"You reckon they could get along with five guys," he joked, grinning at Zach, who had moved to the family room and was sitting in a large, overstuffed chair.

"Depends on who they're carrying," Zach countered.

Caitlin overheard their comments, frowned, and admonished them to be a little more respectful. Zach quickly changed the subject. "You want to go to the machinery sale at Crockett's after the funeral?"

"I do if I get back in time. I'll let you know. We might have time to watch the last couple hours of the sale. That's when they sell the tractors anyway."

Caitlin looked at Zach. "Oh, he'll go for sure. You ought to know that." She knew how much Del liked going to sales and staying current on prices. He subscribed to *Farm Talk*, a weekly regional newspaper and conduit for the sale of just about anything associated with agriculture. He also received several biweekly and monthly farm publications that advertised machinery and cattle. He looked at all of them. Two weeks earlier Del and Zach drove to Oklahoma to get a machine lathe they bought out of *Farm Talk*. They got it home all right, but at over eight thousand pounds, it was so heavy they had to hire a dozer with a crane lift to get it placed in the farm shop. Caitlin wondered when Del would decide he needed a bigger shop.

38

November 17
Little Bit Pregnant

There wasn't much about farming and raising cattle that Del Montgomery didn't like. He enjoyed the planning and the work and the satisfaction of seeing a crop through to harvest, and the husbanding of new batches of spring calves each year. He worried when things went wrong, when the weather turned against him, or when something beyond his control intervened, but he was pragmatic about farming. It wasn't all good times, and there were plenty of risks, but it was rare when he encountered something truly unpleasant. This morning he faced one of those moments.

The day before, Del, Zach, and Harlan had rounded up a herd of cows and penned them up overnight in a holding area adjacent to a corral on the rental farm where Zach and Amber lived. Calves from this herd had been weaned and sold earlier in the month, and the cows were mostly quiet now, no longer bawling for their calves. Del wondered how long the memory of a calf remained with a cow—a few days, a week, maybe longer? Maternal bonds seemed little more than dim memories now, if they remained at all.

This morning, with the cattle penned, Del, Zach, Harlan, and Caitlin awaited the veterinary's arrival to begin pregnancy checks for two herds not worked earlier in the fall. It was an exam with high stakes for cow and cattleman alike. Pregnant cows would be retained another year. Open cows would be culled and sold. Del rarely made exceptions. Last year he had. It was a decision based on sentiment rather than business, and it proved a mistake. A seventeen-year-old cow, a Black Angus show cow that Wendy had entered in competitions at the American Royal in Kansas City and at state fairs when she was in high school, turned up open after the vet's pregnancy check. Del and Caitlin decided to keep her another year, thinking she might breed one more time, even though her age indicated otherwise. It was a sentimental decision, and they knew it. The cow was gentle, and over the years they'd become attached to her. When Del fed this herd, she often came to the door of the truck, even shoved her wet nose against the glass or into the open window, hoping for a handout of molasses-laced supplement. Del often obliged, saving back a handful just for her. But farmers raise livestock for business, not pets or sentiment, and Del was no different in this regard. However, this cow had been special, so they kept her. Several weeks ago she turned up open again when the vet worked the herd. Del and Caitlin knew she'd never raise another calf. Unlike the calves, most of which were sold on Superior's nationwide satellite broadcast auctions on television, Del sold old cows and culls at local auctions. They didn't bring much, half the price of younger bred cows, and professional buyers, who worked these sales like vultures, bought almost all of these cattle for slaughterhouses. Caitlin had told the vet that she wasn't going to watch her sell, and she was pretty sure Del felt the same way.

Even more difficult was the decision Del and Zach faced now. Since last spring, they'd watched a cow with a cancer on her eyelid. She was a Hereford, one of the few they owned, a reddish-brown cow with a white face. Earlier in the year, she had a beautiful calf, and they let her raise it through the summer, even as the cancer spread to the side of her head, a bloody cauliflower that claimed her eye as well. She couldn't be sold—sale barns would reject animals with even the smallest visible defect or health problem—and she probably wouldn't live through the winter. She certainly couldn't raise another calf.

Del and Zach drove down across the small pasture to where the cow was standing. The grass in the pasture was still damp and the air cool. Zach stopped the black flatbed pickup near the cow, which stood alone a quarter mile from the corral where Caitlin and Harlan waited for the vet. She was weak and blind in one eye and stood quietly. Head lowered, she made no attempt to move away, even as the truck approached. The engine went quiet, doors creaked open in the still morning air, and the two men got out. Zach was holding a rifle.

Standing a few feet from the cow, Zach glanced back at Del, then stood rock still. He chambered a bullet, raised the rifle, and took slow, careful aim at an imaginary spot on the cow's skull approximately where the lines of an X would cross if drawn between her eyes and ears. Del leaned against the door of the pickup truck and watched. He said nothing. Overhead a spiral of blackbirds in tight formation spun through a morning sky that still showed faint streaks of orange. Their crackling calls increased, then receded as the flock moved away. Intent on their task at hand, the men took little notice. Mercy killing was one of the most difficult and unpleasant tasks a farmer or cattleman faced. Del had done it many times, knew it had to be done, but it still bothered him. He remembered a story a neighbor had told him years ago. It was about an old man in the community that had grown up farming with horses. Like many of the men of his generation, he was sentimental about his horses. One of the man's horses, a big Belgium long past its working life, got down and couldn't stand, and the old man had asked the neighbor to come and put it down. He said the horse had been a part of his life for so long he couldn't bear to kill it, but he knew the animal was suffering. The neighbor agreed to do it. He'd killed the horse with a single shot to the head, much as Zach was about to do today, but the old man couldn't watch.

Zach flexed his knees slightly, lowering his line of sight closer to that of the cow, who stood passive and uncomprehending. Zach took his time, aware that a slight miscalculation in his aim would only wound the cow and force him to have to shoot again. He didn't want that to happen. He took a deep breath, slowly exhaled, and squeezed the trigger of the black .243 caliber rifle. A popping sound split the air, and the cow reflexively convulsed, her legs and body momentarily stiffening, then abruptly she collapsed. In an instant, with a single bullet to the brain, the animal was dead. Zach lowered the rifle. "That's the first thing this rifle has killed in almost four years," he said softly, as if to break the tension. "This is Amber's deer rifle. She hasn't gotten a deer since her first hunt."

Del nodded. "Well, that's done."

Without further comment, he climbed into the truck and backed it up close to the cow. Zach tied a small chain around her back legs and fastened it to the hitch at the back of the truck. They both knew that if an animal had to be killed, it was best to do it swiftly and in surroundings familiar to the animal. Caitlin watched from the corral, saw Del's truck move away across the pasture, dragging the cow, whose bulbous brown corpse slid easily in the damp grass. Later in the winter, they'd try to burn the dry carcass in a brush pile.

Zach turned and began walking back up the long, sloping pasture towards the corral, rifle in hand. That was when Caitlin heard the vet's truck approaching, pulling up the lane, then saw the truck and red cattle chute pass the house at the top of the hill. She followed the progress of the truck as it turned onto a grassy track in front of a white barn, then lost sight of it behind a large metal shed, but she could hear the engine whining and the steel chute rattling as it bumped along a sagging fence line with volunteer mulberry trees. The vet's truck finally emerged by a gate next to the corral and holding pens. He backed the chute against the corral alleyway, then jumped out, unhooked it, and raised the wheels so it rested on the ground.

Dr. Reid Barkley, a large animal vet, was ruddy in complexion and slight of build, but his boots and hat made him seem larger than he was. He had graying blond hair and blue eyes and an easy manner that seemed to instill calm in the animals that he administered.

The vet's late-model Ford truck had seen too many rough roads and rutted fields. A few ornery cows had already battered character into its once factory-smooth body. However, tucked away between manure-spattered fenders and a bent tailgate was a cornucopia of pharmacological supplies organized inside a near-spotless white metal cabinet that filled the rear of the truck bed. Almost immediately, he dropped the tailgate and began pulling out vaccines, needles, syringes, cleansing alcohol, and containers of dewormer. Zach backed his brown pickup truck parallel to the side of the red chute and began laying out supplies, too, items that they would need—numbered ear tags, an ear tag punch, pliers, scissors, clippers to remove old ear tags, and a log book where each cow's tag number and trimester of pregnancy would be recorded.

By the time Del got back from disposing of the cow, the vet was pulling on waterproof nylon pants, a sleeveless shirt, and a thigh-length waterproof slicker with the left sleeve rolled up to his shoulder. Last was a series of long latex gloves that ran the length of his left arm. He put several on all at the same time. "Saves time," he said, winking at Caitlin, who was watching him. "When one gets dirty, I just peel it off and there's a clean glove underneath."

"If I didn't know better, I'd think you were getting ready for a veterinary fashion show," she said.

Before he could answer, Del glanced at the vet's waterproof gear. "Don't know about any fashion show, but it looks to me like you think it's gonna rain or something."

"Where this arm is going, it'll be plenty wet," the vet retorted with a grin. "Besides, some of these cows of yours got pretty good aim if you know what I mean. I gotta protect myself."

Inside the corral, Harlan, with arms spread, began talking in a soothing tone and moving cows, a few at a time, toward a small wood-fence holding pen that narrowed into a long alley leading to the vet's cattle chute at the far end. "Huu-a," he called. "Git in there, you old gals."

Caitlin, working from outside the corral alley, helped Harlan, talking softly, occasionally prodding a cow forward with her small yellow whip. When the first cow entered the chute, Zach yanked on a lever to release the spring-loaded gate, immobilizing the cow from all sides. Quickly Del closed a second steel gate behind that prevented other cows in line behind from pressing forward while the vet was conducting a pregnancy test.

In one sense, pregnancy checks represented the modern face of veterinary medicine—the ability of a skilled human practitioner to detect a cow's pregnancy with a high degree of accuracy, even before the end of the first trimester when the tiny fetus was too small to be felt physically. Like humans, a fetus in a cow could be detected by ultrasound, but that was expensive and rarely used on ordinary farm animals. Instead, cattlemen opted for a low-tech, less-expensive solution—a physical exam. It usually took less than a minute. Under ideal conditions, a vet might be able to examine a hundred or more cows in a day, but it was physically exhausting work.

Today Dr. Barkley would examine about eighty cows. After each cow was immobilized in the holding chute, the vet inserted a latex-gloved arm to above his elbow into the animal's vagina and, working entirely by touch from inside the cow, felt downward to the uterus and pelvic cradle for evidence signaling a pregnancy and the trimester of development. Early pregnancies were often difficult to verify, and one less than forty-five days along couldn't be reliably diagnosed. During the last half of the first trimester, however, an experienced examiner usually could locate the horns of the uterus and feel the slipping membrane of the embryonic fetus inside one of the slightly enlarged uterine horns. More advanced pregnancies were easier, and the thickened uterus itself, expanding beyond the floor of the pelvic cradle, could be palpated. Sometimes also the hard, nodule-like blood vessels that joined the fetus to the mother's uterus could be felt. The fetus, however, usually couldn't be detected by touch until the third trimester because it was carried so low inside the body cavity.

In another sense this was also the old face of veterinary medicine because it involved a certain amount of physical risk to a vet, who still had to enter the restraining chute with the animal, and one that required a degree of tolerance as regards hygiene. When restrained in the chutes, cows sometimes evacuated their bowels in fright and, under these circumstances, physically inserting a gloved hand and arm into the animal's vagina placed man and animal in intimate but hardly hygienic contact. Rubber gloves and waterproof clothing were no joking matter.

While the vet examined cows and diagnosed pregnancies, Zach, working from the opposite end of the chute, called out ear tag numbers to Del. He also inserted a long, curved tube into each animal's throat and injected a sticky deworming liquid, which had a pasty, chalklike consistency. Judging from bovine reactions to the liquid, it was probably even less appetizing than it looked. Del logged or verified the ear tag number of each cow and poured oily fly repellant onto each animal's back. This late in the season there weren't many flies, but the repellent might help control bot fly larvae, which could still develop under the skin on the backs of cattle. Last were three vaccinations, seven-way blackleg, a five-way Lepto, and a four-way Pasteurella, which every animal that tested pregnant received annually before being released. By the time the last vaccination was administered and the restraining chute opened, every cow knew she'd seen the doctor.

The vet continued to call out his pregnancy results, and Del recorded them in the logbook laying open on the back of the truck. "This one's pregnant. No-o-o problem. Second trimester." His voice was loud and clear above the noise of the animals and the chute.

Pregnancy rates usually averaged eighty-five percent or higher, but Del had noticed variation from year to year, often lower following long, hot summers. "Some guys said almost thirty percent of their cows were open this year. You seeing that?"

"Yep, that's true," the vet said, continuing to talk even as his gloved fingers probed delicately for evidence of new life growing inside a cow. "Hot weather's hard on bulls' sperm counts. Sometimes it's hard to say for sure what the cause is."

"Lot of guys around here been saying it was because of the hot weather we had back in late July and August. They got to blame it on something."

There was a pause, and the vet's face grew serious; he grimaced slightly, his arm almost shoulder-deep inside a cow. His rubber jacket was smeared with manure and urine. He fell quiet, his unseen gloved fingers searching, feeling deep inside the warm innards of the cow for telltale signs of a developing life. Almost a minute passed, an unusually long examination. "Nope! This one's open. Not pregnant. She musta danced all night instead," he finally announced, repeating his findings.

This was the first cow of the morning to turn up open. Del immediately backed the black cattle trailer up to the chute, Zach clipped the cow's ear tag, signaling she would no longer be tracked in the herd, and her number was struck from the active list on the herd registry. Within minutes, she was loaded into the stock trailer. Culled from the herd, her next stop would be a sale barn, although Del sometimes kept open cows penned for a few weeks to get some weight gain before selling them.

The examinations continued without a break—twenty, thirty, forty cows—and sometimes two, even three in a row were open.

"Often like that," the vet said. "It's like all these gals that don't get pregnant hang out together."

"Doesn't pay to keep them. When we did, like with that show cow last year, they usually don't breed again. Then you've got two years of feed and pasture tied up in a non-productive animal."

"This one's pregnant," the vet called out. "First trimester. No-o-o problem," he drawled. "Right now she's a little bit pregnant." Because of the noise and commotion around the corral, he repeated his findings two or three different ways, sometimes with a humorous comment, so there would be no mistakes recording the information.

"What's her tag number?" Del asked.

"Hard to see," Zach said. "She needs a haircut. Got so much hair in her ears I can't read the number."

"Well, here's the scissors, but be careful if she moves her head. Don't want to trim her ear off or else we won't have no place to hang a tag. Here, I'll hold the dewormer tube in her mouth to keep her occupied."

"Okay, I see it. Number ninety-eight, white tag."

"Yeah, she's one of the small ones, but she always has a good calf. Almost as big as she is by the end of summer."

With the next cow the vet paused again. "Open cow," he blurted out. "No! Wait a minute, guys. Nope. She's a little bit pregnant too. First period. Sometimes it's tricky to tell yes or no when it's this early. Hey, eighty-one. You're a lucky gal. Almost took a one-way truck ride to someplace you didn't want to go."

As number eighty-one left the holding chute, Caitlin urged another cow forward, Zach pulled on a chute lever to immobilize her, and Del secured the gate at the rear of the chute. The work was repetitive, but each animal brought its own unique set of needs.

"This one needs a new ear tag," Zach noted.

"Can you read the number?"

"Nope, it's gone."

"All right, here's a new one. We might match her up later to her old number. Don't really matter though."

Suddenly Harlan was cursing loudly in the corral. "Goddamn cow! Get the hell in there," he yelled, slipping as he tried to force a steel gate shut behind a cow and hit his forehead hard against the gate. Caitlin could see a large bruise on his forehead and a trickle of blood down the side of his face.

"It's all right. I'll live," he grumbled, brushing off the accident with a wave of his hand, but he was unusually quiet after the accident.

Caitlin could see he was hurting, but with a cow moving through the chute every few minutes, there was little time for ice packs or medication. She knew that unless it was really serious the men wouldn't stop.

"Here's one with a bare patch on the side of her belly."

"Probably a ringworm," the vet said without looking up.

"Can you treat it?"

"I can, but there isn't any good treatment for it."

"Doesn't iodine work?"

"I'd let it go. Sometimes it just clears up. Besides, the bull didn't care. She's pregnant. Second trimester. No-o-o problem."

"Here's the tame one," Zach said as a large-boned, gray cow with white stockings and a blaze face calmly lumbered into the chute. "She'll eat out of your hand just like Wendy's old show cow."

"I might even let her try again if she's not pregnant," Del said.

"You won't have to do that," the vet answered. "No-o-o problem. She's got a nice calf on the way. Second trimester."

It was past midday before they finished with the eighty cows, sixty-six of them pregnant. More cows were open than Del had hoped, but it could have been worse considering the hot weather during the summer breeding period. Standing beside the chute, the vet peeled off the last of his long rubber gloves and waterproof outerwear and shed his wet, manure-stained rubber pants and jacket in a heap on the ground. Then he put on a clean shirt. He had two small jobs at farms that afternoon but stayed a few minutes to talk, grateful for the break.

He said he'd be back the following week to do tests on Del's last herd, then asked about the cow he'd seen Del dragging off earlier in the morning.

When Del told him what had happened, the vet shrugged. "Had a guy once tell me he'd been deer hunting for several years with a bow. Said he made several hits but never brought one down. I asked him if he tried to track them and he said he hadn't. Probably half of those deer died later from their wounds. At least you didn't try to kill the cow with a bow."

"Happens a lot," Zach said. "Even with rifle hunters. My brother-in-law got a deer here on this farm a couple years ago. Shot it in that brushy area across the road, but it ran and he couldn't find it. It was late, and he looked until dark. Next morning, he went out again and found the deer about a quarter mile away in some brush under a tree in a fence line, but part of it had been eaten during the night. Coyotes had gotten to it."

"You got a place picked out to hunt?" the vet inquired.

"Hopefully a couple. My wife, Amber, she wants to hunt too," Zach said. "She hasn't gotten one since her first try, but she's not giving up. She'll hunt here on the farm. That way she can go out for an hour or two in the evening when she gets home from work. I've got another spot picked out on a neighbor's farm that we rent. He caught some guys turkey hunting there last spring. One of them was a preacher, and he didn't even have a hunting license."

"Don't think I ever heard of a preacher hunting anyway," the vet replied.

"Whatever he was preaching, I don't think he was following his own advice."

Caitlin stayed, too, and listened as the men traded stories about deer hunting and cattle and pickup trucks and farming. It was a beautiful fall day, an Indian summer day, they called it, and in the warmth of the midday sun, it seemed as though autumn could go on forever. Most of the cattle had, by now, wandered out of the holding lot and into an adjacent pasture. The vet finally gathered up his equipment, putting it back in white metal drawers in the back of the truck, and Del and Zach helped him hook the portable cattle chute to his truck. Trees around the farm buildings were still bright with color, reds and oranges on sugar maples, yellows and reds on sweet gums along the creek, and rusts and dark yellows on oaks, but the color wouldn't last, and there would be fewer warm days now. In a distant tree at the edge of a field that was rust and gold with lines of green, a pair of crows perched on a dead limb and eyed the men and women warily. The crows were survivors, consummate opportunists. They would stay for the winter, but winter could bring hard times even for them.

Farming was like that too. It involved uncertainty, and there would always be difficult times. But for a few weeks, Del didn't see any crises on the horizon. Fall cattle chores were almost finished, as was much of the year's seasonal work, so now there might be time to do other jobs. Del wasn't interested in hunting, but he was thinking of shop projects, and he'd finally bought that big articulating tractor that had been in a highway accident. He knew that Zach was eager to start working on it too.

39

November 30
Cow on Fire

November should have been the end of the year. When the calendar year ended in December, farmers, like almost everyone else, might recount their "what ifs" of the past year and begin looking to the promise of the new year, but it was November that brought fiscal closure to most midwestern farms. By then, crops were harvested and either stored or sold, and farmers had put most of their cattle affairs in order as well. With their books effectively closed on the year's most important income-producing activities, farmers would begin to think about the coming year.

They would also begin thinking about taxes. Del and Caitlin, like most farmers, tried to meet with their tax advisor this time of year to make adjustments to debit and credit columns. Professional advice was almost mandatory considering the complexity of farm taxes. Farming was a capital-intensive business, and it was not a coincidence that sales of machinery and cattle were generally well attended this time of year. If the year had been good, it was a sure bet that there would be plenty of pre-ordering of farm supplies, seeds, fertilizer, fencing, and other supplies, as well as cattle purchases, all good tax write-offs before year end.

Caitlin was both relieved and excited to see everything come to a close. "Some years you just hope you come out even or, if you're lucky, a little ahead," she told a friend over the phone. "There may be some money left over after all the bills are paid. We had good harvests this fall."

Overhearing her conversation, Zach turned to Del. "We've still got one more little chore to take care of too. You ready to set some cows on fire?"

"Let's go," Del said, getting up from the kitchen table. "Not gettin' anything done here."

On the back porch they pulled on rubber boots and coveralls. Caitlin was still on the phone. As the porch door slammed behind them, they heard her telling a neighbor that a buyer of some of their calves was the same as last year, a cattle finisher who lived a few hours away. The rest went to a feedlot somewhere in Kansas or Oklahoma.

It wasn't a coincidence that the same buyer bought some of the calves again this year. He liked the quality of the animals and had even stopped by earlier in the fall to look at them. Del was just glad it hadn't rained the day the calves were shipped out to that buyer. The previous weekend it had rained when they loaded out the open cows. "It was so muddy we needed goggles to get anywhere near those cows," Del had said, grinning at Zach.

One truck driver that helped load a batch of calves after a rain had been considerably more graphic in his description of the muddy conditions. Caitlin had overhead him and figured his language might have been a little less colorful in polite company.

Outside, Del and Zach greeted Ross. "Good to get out of the city and breathe some fresh air," he always said. This time he'd come to inspect a barn that needed repair.

"Why don't you stick around a while?" Zach said with a twinkle in his eye. "Wanna see a cow on fire?"

Ross looked nonplussed. He wasn't sure what Zach meant. When they told him they were about to brand some heifers they'd bought, he said he'd never seen a cow being branded except in some western movies. His dad had never done it when he was a kid on the farm.

"A little different now," Del said with amusement.

Branding animals, especially cattle, had a long and illustrious history. In the American West, it was a necessity before there were fences. Even after fences, branding never really disappeared. Ranchers and farmers with large numbers of cattle still branded them, and insurance agents who wrote policies insuring cattle encouraged it. One thing had changed, however. The iconic image of a bunch of cowboys heating a branding iron over an open fire and holding down a calf by hand was gone—long replaced by cattle chutes and electric branding irons.

Zach plugged the branding iron into a wall socket inside the metal-covered barn opposite the house. Harlan guided the first black heifer down the white corral alleyway and into the red holding chute beside the barn. The steel chute banged shut, immobilizing the heifer. Zach, in coveralls and rubber boots, picked up the branding iron. The tip was shaped to form a letter "M" and looked a lot like an oversize potato masher. Zach would apply the brand only on the left side of the heifer and just in front of the rump or round section. A butcher would have called it the loin. With the iron poised an inch above the heifer's upper left rear, he paused momentarily to make sure the iron was properly aligned with the heifer and the heating element right side up.

"Don't want to be putting any Ws on these gals," he said, grinning at Del. "No going back to correct a mistake." Satisfied, he pressed the iron against the heifer and held it for several seconds. There was a sizzling sound and then a billowing mass of white smoke rose from the moisture in the animal's skin and the burning hair. Zach almost disappeared inside the steamy cloud. The heifer protested mildly, squirming a little in the chute, but her subdued response was hardly the reaction expected from a hot iron touching live skin. There was almost no odor from the burning hair.

"Incredible! You'd think they would protest more," Ross exclaimed. "Imagine having that thing on your own butt."

"Some do," Del said. "Others hardly seem to notice. Don't even flinch. Maybe they have thicker hide and don't feel it. Occasionally one kicks and fusses, even tries to lay down in the chute."

"The squirmy ones are the hardest to brand," Zach said. "I have to keep the iron in exactly the same spot. If they're moving around that's hard to do."

After Zach lifted the branding iron and waited for the smoke to clear, a neat imprint of the letter M could be seen where the hair had been burned off and bare skin showed. The heifer's skin was tinged slightly brown, like newly tanned leather.

"Reckon she'll feel a draft there on cold days," Ross said as he stepped back out of the smoke.

"You and I would, but I doubt if she will," Zach replied. "Winter guard hairs pretty much cover up the brand in the winter."

"You ever find a missing cow because of a brand?"

"We have," Del said. "Once a cow was lost for over a month. We didn't expect to see her again. Then somebody saw the brand and called us."

"We've sold branded cattle, too, then seen them show up for sale again at an auction years later," Zach added. "Every brand has to be registered exactly according to its shape and location. Can't just put it anywhere. Get it in the wrong place and then it might be somebody else's brand."

"That'd be kind of like signing a blank check and letting it walk around the pasture," Del added.

"Looks to me like you're grilling a steak every time you brand one," Ross said.

"Well, I like my steaks grilled a little more than that," Del said, laughing.

"You ever try freeze branding?"

"No. I don't know anybody around here that does. It's more complicated having to use dry ice and alcohol. Takes longer to set the brand. I've heard of people using liquid nitrogen, too, but you can hurt an animal with that stuff."

"Hair grows back white?"

"Oh, yeah. It does. Easy to see the brand then, but we've never had any trouble seeing our old-fashioned ones."

It was nearly two hours before the smoke cleared and the last heifer left the chute with a new letter M stamped on her side. There were twenty newly purchased cows and thirty heifers. The job had gone smoothly—Harlan working the corrals and lining up animals, one by one in the alley, Del locking each animal in the chute and releasing them afterward, and Zach doing the branding. It would be the last time they'd have to work cattle for the year, although some years they were still selling cattle in early December. Routine winter feeding chores would start soon. Hay was stockpiled and the feeding trucks were ready.

The afternoon was gray and overcast, the wind chilly. A few juncos and winter sparrows flitted restlessly around the edges of the buildings and corrals and in dry weeds killed by the first freeze of the fall. These hardy little birds always appeared about this time of year. Most of them stayed through the winter, eating seeds in weedy patches. Ducks and geese were on the move now too. Del had seen strings of them in the sky during the past week, some moving south, others ranging out from reservoirs in the evenings to feed on spilled corn in fields.

Del liked this time of the year, the change from fall to winter. Winter brought its share of stress—looking after cattle during calving season, getting around farms when it was muddy or icy, getting trucks and tractors started on cold mornings—but with winter work he didn't feel the financial pressure that came with planting or haying or harvest. His biggest worry was freezing rain and snowstorms during calving time. He liked the rhythm of the seasons, too, and being outside every day. He was also proud that Zach had followed in his footsteps and had reached a point where he could take over the farming business if necessary. Farming was a business that took a lifetime to build, and Del and Caitlin, like every farm couple, looked ahead to the possibility of a child and even a grandchild some day taking over the farming business. Also, Harlan had been with him for over twelve years now, and having a hired hand had enabled him to expand his farming business. He could even foresee employing a second permanent hired hand if his farming business continued to prosper.

Over much of human history, progress in agriculture has been barely incremental at best, century after century often passing with little change. Then, in the middle of the nineteenth century, agriculture was swept up in an industrial revolution that changed forever how humans grew their food and harvested and transported and marketed it. Efficiency and scale increased dramatically, agricultural markets became more international, and farms more specialized and efficient, but it hadn't always been a smooth ride. The Great Depression of the 1930s devastated American agriculture, but enlightened post-Depression policies had raised farmers and agriculture up from their knees again. Del hoped that politicians would continue to stay focused during the coming decades. He thought it would be necessary if family farms were to remain viable.

Del tried to take care of the land he owned, and he took care of his tenants' land too. In the end, it had paid him back—in fields that were more efficient to farm, in crops with better yields, and in cattle that were thrifty and produced good calves. Land ownership, he knew, was temporary, and if a farmer wanted to leave something for his children or grandchildren, or for the future, the land had to be cared for, and that meant an investment of time and effort and money.

Del made most of the major farm decisions, but he valued Caitlin's input. She'd been a good sounding board for his ideas over the year, often asking sharp questions that forced him to consider things he hadn't taken into account. Like most farm wives, she was an integral part of the success of their farming business. Caitlin was comfortable with her role on the farm and in the community. From time to time she might have thought about the "what ifs" had she not stayed on a farm. What if she had pursued another career? Lived somewhere else? Finished her college studies? She knew life was a series of decisions, and each one opened a new door to a potentially different outcome. But she had no regrets, and Del knew that he wouldn't be where he was today without her. They both derived a good deal of satisfaction from shepherding the farm through good times and bad, and from the financial success that it had brought them.

Del stood by the corrals looking at his newly branded heifers and at spare pastures and fields that had recently begun to take on the earthtones of winter. He was thinking about the year coming to a close when Zach walked up.

"Hey, Dad, Charlie just called and said he saw some of our cows out in the road over by his place. He thought they might have gotten out last night and that we'd want to know."

"We certainly do. Let's go get those gals back in and fix the fence. I guess I wasn't planning on taking the rest of the afternoon off anyway."

EPILOGUE

Over the following year, some co-ops quit supplying anhydrous ammonia to farmers because of liability issues resulting from anhydrous ammonia theft for the production of methamphetamine. Del's nearest supply of anhydrous ammonia was now farther away, and to adjust to this inconvenience, he had purchased a twin-tank trailer that could be towed at highway speed. Because of the recent increase in commodity prices, Del had begun growing wheat again and, whenever possible, had been double-cropping soybeans into wheat stubble immediately after the mid-June harvest. He had also bought a newer and even larger no-till planter. With commodity price increases, he had expanded acreage of both corn and soybeans and had, for now, eliminated milo from his crop rotation. He remained acutely aware, however, that the mix of small grain and row-crop acreage that he planted each year would have to be continually adjusted to market conditions. He also noted that Roundup, a staple herbicide in American agriculture, had gone off-patent and, coincidentally or not, no longer seemed as effective.

For some time, Harlan had been talking about moving west to work construction, but things never seemed to work out. For a while he lived in a small travel trailer, which he parked on a friend's property. It was close to an arm of a lake where he could fish during his free time. After almost fourteen years he finally left his job with Del, fashioned a reconciliation with his wife, and was happily tending a large vegetable garden. He continues to fish whenever possible. Del had hired another farmhand and also a second hired hand during busy times. Now both are permanent employees.

Zach and Amber had a second baby boy born in June of the following year. Caitlin continued working part time at the post office, but her job was eventually discontinued, which suited her fine. She bought a horse, but when it became difficult to handle, Del worried that she would get hurt. He sold the horse and bought another, which proved easier to handle. Caitlin continues to saddle and ride occasionally, and they also bought a Shetland pony and small cart, which has been popular with the grandchildren.

After finishing his pulling tractor, Zach then built another even more powerful one. His new tractor was quite successful, and after one pull his son Nathaniel said, "Daddy did good. He didn't blow anything up." As his family has grown, Zach has reduced his involvement in tractor pulling in favor of spending more time with his family, and especially with his boys' scouting activities.

After Maggie the orphan dog left Zach and Amber for a neighbor, her life didn't gain any stability. The neighbor's home burned, forcing him to move, and Maggie got left behind. Her present whereabouts are unknown. Crissy, the Montgomerys' black and white herding dog, fared even less well, first enduring an injury beneath the wheels of an Amish buggy, and some months later suffering a fatal kick from the Montgomerys' horse. On the other hand, Boy the shop cat still lives comfortably in the farm shop and is now twenty years old. The house and porch cats remain at their respective posts as well. One is even older than Boy.

After Del and Zach purchased the damaged eight-wheel drive, center-articulating tractor, they replaced an axle, made numerous repairs, and had the huge tractor ready for fieldwork the following year. They liked it so much that they have since bought two more. The Montgomerys continue to expand their farming operation, adding more trucks, newer and larger farm equipment, and a recent land purchase. And, as Caitlin predicted, Del did enlarge his farm shop. Weather remains a variable, as always, but crop prices have increased in recent years and cattle prices, always cyclical, are now much higher. Del and Caitlin remain optimistic about the future of family farms and about their joint farming venture with Zach. Looking back on forty years of hard work, and on their relatively modest beginnings, they seem a little surprised at the success of their farming operation. It is one among a gradually declining number of family farms that has steadily prospered. They have every right to be proud of their accomplishments.